MACROERGONOMICS

AN INTRODUCTION TO WORK SYSTEM DESIGN

-- -- -- -- -- -- -- --> by Hal W. Hendrick and Brian M. Kleiner

HFES Issues in Human Factors and Ergonomics Book Series, Volume 2
Supervising Editor: Thomas J. Smith

Published by the
Human Factors and Ergonomics Society
P.O. Box 1369
Santa Monica, CA 90406-1369 USA
310/394-1811, Fax 310/394-2410
http://hfes.org, info@hfes.org

Permission to reproduce any chapter or a substantial portion (more than 300 words) thereof, or any figure or table, must come from the author and from the HFES Publications Department. Reproduction or systematic or multiple reproduction of any material in this book is permitted only under license from the Human Factors and Ergonomics Society. Address inquiries to the Communications Department, Human Factors and Ergonomics Society, P.O. Box 1369, Santa Monica, CA 90406-1369 USA; 310/394-1811, fax 310/394-2410, e-mail lois@hfes.org.

Additional copies of this book may be obtained from the Human Factors and Ergonomics Society for $35.00 per copy for members and $40.00 per copy for nonmembers. Add $7.00 for shipping and handling and California sales tax if applicable. Discounts apply on purchases of five copies or more; contact the Society at the address given above.

The Human Factors and Ergonomics Society is a multidisciplinary professional association of 5000 persons in the United States and throughout the world. Its members include psychologists, designers, and scientists, all of whom have a common interest in designing systems and equipment to be safe and effective for the people who operate and maintain them.

Library of Congress Cataloging-in-Publication Data

Hendrick, Hal W.
    Macroergonomics : an introduction to work system design /
Hal W. Hendrick and Brian M. Kleiner.
        p. cm. — (HFES issues in human factors and
ergonomics book series : volume 2)
        Includes bibliographical references and index.
        ISBN 0-945289-14-6
        1. Human engineering. I. Kleiner, Brian M., 1959- II. Title.
III. Series.
        TA166 .H36 2000
        620.8'2–dc21   00-012084
                                                          CIP

CONTENTS

Chapter 3. COMMON MACROERGONOMICS METHODS   30

Chapter 4. ANALYSIS AND DESIGN OF WORK SYSTEM STRUCTURE   47

Chapter 5. ANALYSIS AND DESIGN OF WORK SYSTEM PROCESS   67

Chapter 6. MACROERGONOMICS RESULTS   88

PREFACE

This book reports on the development of the subdiscipline of human factors and ergonomics that has come to be known as macroergonomics. It represents a significant milestone in the development of macroergonomics: Many journal and proceedings articles and book chapters have been written on this topic over the past 20 years, but a concise, introductory book has been lacking – until now.

Macroergonomics: An Introduction to Work System Design is for those practitioners, managers, scholars, and students who desire a basic introduction to the key concepts and issues in macroergonomics. We view macroergonomics from a value-added perspective as a recognized subdiscipline of ergonomics, as a methodology, and as an empirical science. Whether you are a macroergonomist seeking a core resource on the discipline or an ergonomist or student seeking a broader perspective, you'll find much of value in this book.

At this time in our profession, when ergonomics needs greater appreciation and acceptance, we are hopeful that the larger system perspective of macroergonomics can help raise awareness that ergonomics is a relevant and cost-effective science and practice that can improve the human condition.

Chapter 1

INTRODUCTION

This book reports on the development and characteristics of the subdiscipline of human factors or ergonomics that has come to be known as *macroergonomics*. Macroergonomics deals with the analysis and design of *work systems*. As used here, work is used generically to refer to any form of human effort or activity. *System* refers to *sociotechnical* systems (see Chapter 2). These systems may be as simple as a single individual using a hand implement or as complex as a multinational organization. A *work system* is one that involves two or more people interacting with some form of (a) hardware and/or software, (b) internal environment, (c) external environment, and/or (d) an organizational design.

The hardware typically consists of machines or tools. The internal environment consists of various physical parameters, such as temperature, humidity, illumination, noise, air quality, and vibration. It also includes *psychosocial factors* (see Chapter 3). The external environment consists of those elements that permeate the organization and to which the organization must be responsive in order to survive and be successful. Included are political, cultural, and customer attributes; materials and parts resources, available labor pool; and educational resources. The organizational design of a work system is composed of both an organizational structure and the processes by which the work system accomplishes its functions.

Included in this overview is the historical development and description of macroergonomics, its major theoretical constructs, the most commonly used methods and tools, and some laboratory findings and case studies that provide validation of its constructs. Finally, some work in progress and developmental trends are noted.

Historical Development

Since its formal inception in the late 1940s, human factors or ergonomics (the discipline is known by both terms, sometimes in combination) has been concerned with designing sociotechnical systems to optimize people's interaction with systems, tools, products, and environments. Prior to the 1980s, the focus of this design concern was on optimizing interactions between operators and their work environment, or what is often referred to as *micro-*

ergonomics. This focus initially was labeled *man-machine* interface design (McCormick, 1957). In the 1970s, with growing sensitivity to gender issues, the term *human-machine* interface design came into use. The objective was to apply scientific knowledge about human capabilities, limitations, and other characteristics to the design of operator controls, displays, tools, workspace arrangements, and physical environments to enhance health, safety, comfort, and productivity and to minimize human error through design.

With the advent of the silicon chip and the subsequent rapid development of computers and automation, a new subdiscipline of human factors/ergonomics (HF/E) emerged. It centered on software design and became known as *cognitive ergonomics* (Dray, 1985, 1986). The focus was still on the individual operator but with an emphasis on how humans think and process information and how to design software to communicate in the same manner.

The Beginning: Select Committee on Human Factors Futures

In 1978, former Human Factors Society (HFS) President Arnold Small noted that many dramatic changes were occurring in all aspects of society and its built environment. Small believed that traditional human factors/ergonomics would not be adequate to respond to these trends effectively. At his urging, HFS (renamed the Human Factors and Ergonomics Society in 1993) formed the Select Committee on Human Factors Futures, 1980–2000, to study these trends and determine their implications for the HF/E discipline. Small was appointed chair. One of the authors of this book, Hal Hendrick, was appointed to the committee and charged with researching trends related to the management and organization of work systems.

In October 1980, at the HFS Annual Meeting in Los Angeles, California, the committee members reported their findings. Among other things, Hendrick (1980) noted the following six major trends:

1. *Technology.* Breakthroughs in the development of new materials, microminiaturization of components, and the rapid development of new technology in the computer and telecommunications industries would fundamentally alter the nature of work in offices and factories during the period 1980–2000. In general, society had entered a true information age of automation that would profoundly affect work organization and related human-machine interfaces.

2. *Demographic shifts.* The average age of work populations in the world's industrialized countries would increase by approximately six months for each year during the 1980s and most of the 1990s. Two major factors account for this "graying" of the workforce: (1) the aging of the post-World War II "baby boom" demographic bulge that had entered the workforce and (2) the lengthening of workers' productive life span because of better nutri-

tion and health care. In short, during those two decades, the workforce would become progressively more mature, experienced, and professionalized.

As the organizational literature had shown (e.g., see Robbins, 1983, for a more recent review), as the level of professionalism (i.e., education, training, and experience) increases, it becomes important for (a) work systems to become less formalized (less controlled by standardized procedures, rules, and detailed job descriptions), (b) tactical decision-making to become decentralized (delegated to the lower-level supervisors and workers), and (c) management systems to similarly accommodate. These requirements represented profound changes to traditional bureaucratic work systems and related human-system interfaces.

3. *Value changes.* Beginning in the mid-1960s and progressing into the 1970s, a fundamental shift occurred in the value systems of workforces in the United States and Western Europe. These changes and their implications for work systems design were noted by a number of prominent organizational behavior researchers and were summarized by Argyris (1971). In particular, Argyris noted that workers now both valued and expected to have greater control over the planning and pacing of their work, greater decision-making responsibility, and more broadly defined jobs that enable a greater sense of both responsibility and accomplishment. In addition, Argyris stated, to the extent organizations and work system designs did not accommodate these values, organizational efficiency and quality of performance would deteriorate.

These value changes were further validated in the 1970s by Yankelovich (1979) based on extensive longitudinal studies of workforce attitudes and values in the United States. Yankelovich found these changes to be particularly dramatic and strong among workers born after World War II. Of particular note from his findings was the insistence that jobs become less depersonalized and more meaningful.

4. *World competition.* Progressively, U.S. industries were being forced to compete with the influx of high-quality products from Europe and Japan; other competition would soon follow from countries such as Taiwan and Korea. Put simply, the post-World War II dominance by U.S. industry no longer existed. In light of this increasingly competitive world market, the future survival of most companies would depend on their efficiency of operation and production of state-of-the-art products of high quality. In the final analysis, the primary difference between successful and unsuccessful competitors would be the quality of their products and of their total work organization, and the two were likely to be interrelated.

5. *Ergonomics-based litigation.* In the United States, litigation based on the lack of ergonomics safety design considerations in consumer products and

the workplace was increasing, and awards by juries were often high. The message from this litigation was clear: Managers *are* responsible for ensuring that adequate attention is given to the ergonomic design of both their products and their employees' work environments to ensure safety.

One impact of this message was that ergonomists were likely to find themselves functioning as true management consultants. A related implication of equal importance was that ergonomics education programs would need to provide academic courses in organizational theory, behavior, and management to prepare their students for this consultant role.

6. *Failure of traditional (micro-)ergonomics.* Early attempts to incorporate ergonomics into the design of computer workstations and software had resulted in improvement, but the results had been disappointing in terms of reducing the work system productivity costs of white-collar jobs, improving intrinsic job satisfaction, and reducing symptoms of high job stress. It is not that traditional micro-ergonomics had failed to be effective. Indeed, its application had frequently resulted in major improvements in health, safety, and productivity. Rather, it was that by itself, traditional micro-ergonomics had been inadequate for fully accomplishing these goals.

As Hendrick noted several years later (1984, 1986a, 1986b), human factors/ergonomics professionals had begun to realize that it was entirely possible to do an outstanding job of ergonomically designing a system's components, modules, and subsystems but fail to reach relevant systems effectiveness goals because of inattention to the *macro*-ergonomic design of the overall work system. Similar conclusions have been drawn from investigations by Meshkati (1986) and Meshkati and Robertson (1986) of failed technology transfer projects, by Meshkati (1991) of major system disasters (e.g., Three Mile Island and Chernobyl nuclear power plants and the Bhopal chemical plant), and by Munipov (1990) of the Chernobyl accident.

Integrating Ergonomics with ODAM

Based on the foregoing observations, Hendrick concluded in his 1980 report that for the human factors/ergonomics profession to be truly effective and responsive to the foreseeable requirements of the next two decades and beyond, there was a strong need to integrate organizational design and management (ODAM) factors into research and practice.

It is interesting to note that all the predictions from 1980 have come to pass – and are continuing. We believe that these trends also account for the rapid growth and development of macroergonomics that has occurred. In 1984, as a direct response to Hendrick's report, an ODAM technical group was formed within the Human Factors Society, and similar groups were formed within Japan's Ergonomics Research Society (now the Japan Ergonomics Society) and the Hungarian ergonomics society; less formal interest groups were formed in other ergonomics societies internationally. In 1985, the International Ergonomics Association (IEA) formed a Science and Technology Committee comprising eight technical groups, one of which was the ODAM Technical Group (TG). This TG consistently has been one of the IEA's most active, particularly in organizing highly successful IEA international symposia on human factors in ODAM.

By 1986, sufficient conceptualization of the ergonomics of work systems had been developed to identify it as a separate subdiscipline, which became formally identified as *macroergonomics* (Hendrick, 1986a, 1986b). In 1998, as a result of the considerable methodology, research findings, and practice experience that had developed during the 1980s and 1990s, the HFES ODAM Technical Group changed its name to the Macroergonomics Technical Group.

In 1988, ODAM was one of the five major themes of the Xth IEA Triennial Congress in Sydney, Australia. It was one of 12 themes for the XIth Triennial Congress in Paris in 1991. At both the XIIth Congress in Toronto in 1994 and the XIIIth Congress in Tampere in 1997, a major multisession symposium on human factors in ODAM was organized. For the two latter congresses, more papers on macroergonomics and ODAM were received than on any other topic.

The Concept of Macroergonomics

One way to define any scientific discipline is by the nature of its unique technology. As a result of its survey of human factors/ergonomics internationally, in 1996, the HFES Strategic Planning Committee identified the unique technology of human factors/ergonomics as *human-system interface technology*. Interaction occurs between the people within systems and the other sociotechnical system components, including hardware, software, environments, jobs, and work system structures and processes. In the *science* of human factors/ergonomics, researchers study human performance, capabilities, limitations, and other characteristics and use this knowledge to develop human-system interface technology. This technology takes the form of interface design principles, guidelines, specifications, methods, and tools.

In the *practice* of human factors/ergonomics, professionals apply human-system interface technology to the design, analysis, test and evaluation, standardization, and control of systems. The overall goal of the discipline is to improve the human condition, including health, safety, comfort, productivity, and quality of life (Human Factors and Ergonomics Society, 1998).

Human-system interface technology has at least five clearly identifiable subparts, each with a related design focus (Hendrick, 1998):

1. Human-*machine* interface technology, or hardware ergonomics
2. Human-*environment* interface technology, or environmental ergonomics
3. Human-*software* interface technology, or cognitive ergonomics
4. Human-*job* interface technology, or work design ergonomics
5. Human-*organization* interface technology, or macroergonomics

The first four of these technologies focus primarily on the individual or subsystem level. They thus constitute the technologies of micro-ergonomics. The fifth focuses on the overall work system level and thus is the primary technology of macroergonomics.

What Is Macroergonomics?

Is macroergonomics a subdiscipline, an empirical science, a methodology, or a perspective? The simple answer is yes, yes, yes, yes! As its history suggests, macroergonomics is a recognized subdiscipline of human factors/ ergonomics. It also is backed by empirical science. From its foundational research roots in the sociotechnical systems tradition to modern laboratory investigation of the relationships among technological, personnel, organizational design, and environmental variables and their interactions, new scientific knowledge about work systems and work system design has emerged. Complementing the empirical research, systematic macroergonomics methodologies for analysis and design of work systems have emerged as well. And for the ergonomics generalist or micro-ergonomics specialist, macroergonomics represents a perspective that provides the ergonomist with an appreciation for the larger system – a perspective that will increase the likelihood that micro-ergonomics interventions will succeed.

In summary, as a subdiscipline, macroergonomics is concerned with human-organization interface technology. The empirical science supporting this subdiscipline is concerned with factors in the technological subsystem, personnel subsystem, external environment, organizational design, and with their interactions and is guided by sociotechnical systems theory. This will be further discussed in Chapter 4. The subdiscipline also has organizational improvement methodologies, such as sociotechnical analysis of work system structure and sociotechnical analysis and design of work system process (to be described in Chapters 4 and 5). As a perspective, macroergonomics provides certain guiding principles to aid the ergonomist, including participation, flexibility, joint optimization, joint design, continuous improvement of processes, and system harmonization.

> *Conceptually*, macroergonomics is a *top-down* sociotechnical systems approach to the design of work systems and the application of the overall work system design to the design of the human-job, human-machine, and human-software interfaces.

Implementing Macroergonomics

Macroergonomics is top-down conceptually because in the final analysis, the macroergonomist must ensure that the overall work system design is compatible with the organization's sociotechnical system characteristics and that the design of the subunits and components of the work system harmonize with the overall design. However, in practice, the macroergonomics design process is *top-down, bottom-up, and middle-out.* In other words, structures and processes that constitute the overall work system can be analyzed and designed starting (a) with the overall work system structure and processes, and then working down through the system's subsystems and components; (b) with the components, and systematically building up to the overall work system structures and processes; or (c) at an intermediate level in the organization, and systematically building both up and down. Most often, a combination of all three strategies is used, and the process frequently involves employee participation at all levels of the organization.

Furthermore, the macroergonomics design process is often iterative (design, evaluate, refine, reevaluate, further refine, etc.), nonlinear (does not proceed in a simple sequential manner), and stochastic (requires making inferences or decisions based on incomplete data). Only rarely is macroergonomics design a pure process. For example, one often has to accept some existing subsystems or components. Union-management contracts, ongoing projects, and aspects of corporate culture are some of the factors that may prevent implementing what appears to be optimal from a macroergonomics design standpoint. In some cases, parts of the macroergonomics design may simply be delayed in implementation. In other cases, desired changes may never be possible.

A full macroergonomics effort is most feasible when a major work system change is already scheduled to take place, such as changing to a new technology, replacing equipment, or moving to a new facility. Another opportunity exists when there is a major change in the goals, scope, or direction of the organization – for example, when a manufacturing company decides to expand into a new product line or go from being a mass producer to making highly customized products.

Managers may also be receptive when the organization has a costly chronic problem that has not proven correctable with a purely micro-ergonomics effort or through other intervention strategies. Recently, the desire to reduce lost-time accidents and injuries and related costs has led senior managers in some organizations to support a more comprehensive macroergonomics intervention. As illustrated in Chapter 6, many of these efforts have achieved dramatic results.

Frequently, a macroergonomics change to the work system is not possible initially. The ergonomist or ergonomics team begins by making micro-ergonomics improvements that yield positive results in a relatively short time. When managers see these positive results, they become interested in and willing to support further ergonomics interventions. In this process, if the ergonomist or ergonomics team has established a good rapport with the key decision makers, decision makers' level of awareness is raised about the full scope of ergonomics and its potential value to the organization. Over time, senior managers come to support progressively larger ergonomics projects – ones that change the nature of the work system as a whole. Based on the experience and observations of one of the authors, this process typically takes about two years from the time one has established the necessary rapport and gained the confidence of the key decision maker(s).

Macroergonomics versus Industrial/Organizational Psychology

In a sense, industrial and organizational (I/O) psychology is the opposite side of the coin from ergonomics. Whereas ergonomics focuses on designing work systems to fit people, I/O psychology primarily is concerned with selecting people to fit work systems. This is particularly true of classical industrial psychology, which is the opposite of micro-ergonomics: Whereas micro-ergonomics focuses on designing jobs, work environments, hardware, and software to fit individuals, classical industrial psychology focuses on identifying and selecting people to fit jobs.

Nevertheless, there is a greater overlap between organizational psychology and macroergonomics than between classical industrial psychology and micro-ergonomics. The former are both concerned with the design of organizational structures and processes, but the focus differs. In organizational psychology, fostering teamwork, enhancing leadership, and improving motivation, job satisfaction, and incentive systems are common objectives. Although these objectives are also important in macroergonomics, the primary focus of macroergonomics is to design work systems that are compatible with an organization's sociotechnical system characteristics and then to carry that work system design through to the design of human-job, human-machine, human-software, and human-environment interfaces to ensure a fully *harmonized work system*.

Because there is an overlap of macroergonomics with organizational psychology, a number of the empirically developed methods and tools of organizational psychology – and the supporting literature base - are useful in implementing the macroergonomics process. Some critics have claimed that macroergonomics is merely an extension of organizational science to ergonomics and not really ergonomics at all. In fact, because ergonomics centers on the scientific study of human behavior and the application of that knowledge to the design of human-system interfaces in organizations (among other things), one could argue that in at least some respects, ergonomics *is* an organizational science, at both the macro and the micro level.

Nevertheless, our response to the aforementioned criticism is that macroergonomics is distinct from the traditional organizational sciences because it focuses on designing work systems to be compatible with their sociotechnical system characteristics and on carrying that design through to ensure optimal human interaction with jobs, machines, and systems. The organizational sciences, by contrast, focus on ways to improve organizational effectiveness by improving incentive systems, teamwork, leadership, and the organizational climate, among other things.

Pitfalls of Traditional Approaches to Work System Design

Over a 20-year period, one of the authors was involved in assessing more than 200 organizational units. Based on these assessments, he has identified three highly interrelated work system design practices that frequently lead to dysfunctional attempts to develop and modify work systems: (a) technology-centered design, (b) a "leftover" approach to function and task allocation, and (c) a failure to consider an organization's sociotechnical characteristics and integrate them into its work system design (Hendrick, 1995). A description of Hendrick's findings and conclusions follows.

Technology-Centered Design
Designers incorporate technology into some form of hardware or software to achieve some desired purpose. The designer focuses initially on functionality – what the machine can do – and then worries about human functions. Usually the extent to which those who must operate or maintain the hardware or software are considered accounts for the skills, knowledge, and training that will be required. However, these factors are not always considered from an ergonomics standpoint. As a result, the intrinsic motivational aspects of jobs, psychosocial characteristics of the workforce, and other related work

system design factors rarely are considered – yet these are the very factors that can significantly improve work system effectiveness.

Although there are many good examples of early involvement of ergonomists in the design process, from our experiences they constitute a distinct minority of design efforts. More often, if ergonomics aspects of design are considered, it is *after* the equipment or software is designed. Then the ergonomist may be called in to modify some of the human-system interfaces to reduce the likelihood of human error, eliminate awkward postures, or improve comfort. Even this level of involvement sometimes does not occur until testing of the newly designed system reveals serious interface design problems. At this point in the design process, because of cost and schedule considerations, the ergonomist is severely limited in making fundamental changes to improve the work system. Instead, he or she can make only a few "Band-Aid" fixes of specific human-machine, human-environment, or human-software interfaces. Ultimately, the outcome is a work system that functions well below what otherwise would be possible.

There is a widely acknowledged relationship between ergonomics input to design and level of system performance: The earlier the input of professional ergonomists in the design process, the greater the impact on system effectiveness. When employees are not actively involved throughout the planning and implementation processes, the result is often a poorly designed work system and a lack of employee commitment. Frequently, employees even display overt or passive-aggressive resistance to the changes.

From our observations, when a technology-centered approach is taken, if employees are brought into the process at all, it is only after the work system changes have been designed, and employees' role is merely to conduct a usability test. When employees find serious problems with the changes (as often happens), cost and schedule considerations prevent any major redesign to eliminate or minimize the deficiencies.

Given that most of the so-called reengineering efforts of the early 1990s used a technology-centered approach, it is not surprising that most of them have been unsuccessful (e.g., Keidel, 1994). As Keidel noted, these efforts failed to address the "soft" (i.e., human) side of engineering and often ignored organizational effects. For example, one of the authors is familiar with a case involving a large telecommunications company. A reengineering project was implemented to streamline a complex work process. Central to the effort was developing a large new software package. During development, the focus was on getting the software to work. Only after the software package was developed were employees brought into the process. Although on a technical level the software worked, employees found that it actually degraded work performance. It was not compatible with the work system and resulted in less intrinsically motivating jobs than did the old process. The employees were readily able to determine the deficiencies of the software and to make recommendations as to how it could have been designed to streamline their work processes and enhance intrinsic job motivation. By then, however, the cost of a major redesign of the software and schedule considerations prevented implementation of most of the employees' recommendations. The system was implemented, and about a year later, it had to be scrapped.

"Leftover" Approach to Function and Task Allocation

A technology-centered approach often leads to treating those who will operate and maintain the system as impersonal components. The focus is on assigning to the "machine" any functions or tasks that its technology enables it to perform. What is left over is assigned to the operators and maintainers. As a result, the function and task allocation process fails to consider the characteristics of the workforce and related external environmental factors. Typically, the consequence is a poorly designed work system that fails to make effective use of its human resources.

As we discuss in greater detail in Chapter 2, effective work system design requires *joint design* of the technical and personnel subsystems (DeGreene, 1973). In ergonomics terms, joint design requires a *human-centered approach*. In terms of function and task allocation, Bailey (1989) refers to it as a *humanized task* approach:

> This concept essentially means that the ultimate concern is to design a job that *justifies* using a person, rather than a job that merely can be done by a human. With this approach, functions are allocated and the resulting tasks are designed to make full use of human skills and to compensate for human limitations. The nature of the work itself should lend itself to internal motivational influences. The left over functions are allocated to computers. (P. 190)

Failure to Consider the System's Sociotechnical Characteristics

The primary structural and process characteristics of sociotechnical systems were empirically identified first by the Tavistock Institute in the United Kingdom based on studies of long-wall coal mining and other industries in the 1940s and 1950s (DeGreene, 1973). From the literature, four major characteristics or elements of sociotechnical systems can be identified: (a) technological subsystem, (b) personnel subsystem, (c) external environment, and (d) organizational design. These elements interact, so a change in any one affects the other three (often in dysfunctional or unanticipated ways).

Because of these interrelationships, characteristics of each of the first three elements affect the fourth, the organizational design of the work system. In

Chapter 4, we describe these elements and empirical models of these relationships. These models can be used to determine the optimal work system structure (i.e., the optimal degrees of vertical and horizontal differentiation, integration, formalization, and centralization to design into the work system; see Chapter 4).

Unfortunately, as documented in the Tavistock studies of coal mining more than four decades ago (Emery & Trist, 1960; Trist & Bamforth, 1951), a technology-centered approach to the organizational design of work systems does *not* adequately consider the key characteristics of the other three sociotechnical system elements. Consequently, the resulting work system design is usually suboptimal.

Criteria for an Effective Work System Design Approach

Based on the pitfalls just noted, several criteria can be established for selecting an effective work system design approach.

1. *Joint design.* The approach should be *human centered.* Rather than designing the technological subsystem and requiring the personnel subsystem to conform to it, the approach should require design of both subsystems concurrently. Further, it should allow for extensive employee participation throughout the design process.

2. *Humanized task approach.* The function and task allocation process should first consider whether there is a need for a human to perform a given function or task before allocating functions to either humans or machines. Implicit in this criterion is a systematic consideration of the professionalism (education and training), cultural, and psychosocial characteristics of the personnel subsystem.

3. *Consider the organization's sociotechnical characteristics.* The approach should systematically evaluate the organization's sociotechnical system characteristics, and then integrate them into the work system's design.

Macroergonomics fulfills all three criteria because, as noted earlier, it is a top-down sociotechnical systems approach to work system design and the carry-through of the overall work system design to the design of human-job, human-machine, and human-software interfaces. As we describe in Chapters 3 and 4, it is top-down in that it begins with an analysis of the relevant sociotechnical system variables and then systematically uses these data in designing the work system's structure and related processes.

Macroergonomics is human-centered because it systematically considers the worker's professional and psychosocial characteristics in designing the work system and then carries the work system design through to the ergonomic design of specific jobs and related hardware and software interfaces. Integral to this human-centered design process is joint design of the technical and personnel subsystems, using a humanized task approach in allocating functions and tasks. A primary methodology of macroergonomics – and one that many macroergonomics practitioners consider necessary to ensure success – is *participatory ergonomics* (Noro & Imada, 1991). Participatory ergonomics is a methodology that involves employees at all organizational levels in the design process (including function and task allocation) and will be described further in Chapter 3.

Determining the Cost-Benefits of Macroergonomics Interventions

Regardless of other benefits that may be realized from macroergonomics improvements to the work system, organizations usually are not able to justify the intervention unless there is a clear economic benefit to doing so. Accordingly, in developing a macroergonomics intervention proposal for managers, it is extremely important to clearly identify the costs and economic benefits that can be expected and outline how they will be measured. The good news is that properly planned and implemented macroergonomics projects usually do result in significant economic benefits. A number of case studies supporting this statement are provided in Chapter 6.

The following sections describe some of the common costs and benefits to consider in developing a macroergonomics intervention proposal.

Determining the Costs

Although the process can be complex, measuring the costs of macroergonomics projects is usually easier than measuring the benefits. This is because often the cost factors are fewer in number and the necessary accounting data are already available within the organization. For most macroergonomics projects, there are four major classes of costs to consider: (a) personnel, (b) equipment and materials, (c) reduced productivity or sales, and (d) overhead.

Personnel. The cost of hiring a macroergonomics consultant or consulting group comes directly from the consultant's schedule of rates and projected time commitment for the project or from a contract proposal. If the project is to be carried out by one or more internal professionals, then their salary, benefits, and overhead are simply prorated based on their time commitment. This is also true for other in-house personnel who are to work on the project.

If the project will require modifications that will result in employees' not being able to perform their normal work during the project, their "down time" also needs to be factored into the cost. Often, during this down time, employees are put to work on other tasks, or the time is used for required training; in such cases, the down time should not be charged against the project.

Not infrequently, if the macroergonomics intervention is being carried out as part of a changeover of equipment or product line, remodeling, or move to a new facility, the down time would have resulted anyway. In these cases, only those additional costs resulting from the macroergonomics intervention per se should be considered.

Equipment and materials. Because most equipment and materials are purchased directly, the actual purchase costs can be used directly. Note that equipment costs can be treated either as one-time (capital) or as continuing costs (which considers the life of the application). For example, a new tool might cost $300 to purchase, but it could be charged at $50 per year for six years. If only some years of use are directly related to the project, then only the amount for those years should be charged. If equipment is purchased with money from a loan, the interest charges also must be included.

For parts or equipment that are fabricated internally, costs can be determined using the company's cost accounting data. Similarly, if there are storage charges associated with the equipment or materials, either the storage rate for external storage or the company's internal expense rate for the storage space can be used. But again, those equipment and materials costs and related installation or storage expenses that would have been incurred if there were no macroergonomics intervention should *not* be included.

Sometimes the equipment being replaced is either resold or used elsewhere in the company. If this equipment would not have been replaced except for the macroergonomics intervention, its resale price or book value should be *credited* to the intervention project.

Although this is rare, macroergonomics interventions can result in increased maintenance costs above those that otherwise would have been incurred. When this happens, these additional costs also need to be charged against the project. In most cases, macroergonomics interventions actually *reduce* maintenance costs and thus show up as a benefit.

Reduced productivity or sales. Macroergonomics interventions may temporarily disrupt ongoing operations, resulting in a reduction in productivity or sales for a time. The cost of this lost revenue also needs to be considered.

Overhead. Overhead costs, such as facilities maintenance and general administration, typically are calculated by the organization's accounting department and then applied as a percentage of direct costs. Sometimes a macroergonomics intervention may reduce some of these costs. Under these circumstances, the accounting department should be requested to reassess the unit's overhead rate.

Determining the Personnel-Related Benefits

Although it can sometimes take more effort than figuring the costs, determining many of the financial benefits of macroergonomics interventions is easier than might at first appear. Benefits fall into two general economic classes: those associated with *personnel* and those associated with *materials and equipment*. Personnel benefits include (a) increased output per worker; (b) reduced accidents, injuries, and illness; (c) reduced training time; (d) reduced skill requirements; (e) reduced maintenance time; and (f) reduced absenteeism.

Increased output per worker. Improvements to the work system structure and processes can often result in a major benefit: greater output per worker. This can be calculated in terms of the labor cost of each additional item produced per worker. If the output is a service, the economic benefit of the increased service can be calculated in terms of the charge per hour for that additional service provided.

Reduced accidents, injuries, and illness. This is one of the most frequently encountered benefits of macroergonomics interventions in production and maintenance organizations. A common measure is the reduction in lost time from accidents, injuries, and illness. These costs can be multiplied by the labor cost per unit of time to determine the economic benefit. Alternatively, the economic benefit may be the savings in workers' compensation insurance premiums that result.

Reduced training time. Reductions in training requirements may come about because work system changes result in functions and processes that require less time to learn and are easier to perform. Alternatively, training requirements may be reduced because of (a) less turnover, (b) reductions in lost time from accidents and injuries, (c) less absenteeism, or (d) reduced number of people needed to perform a given function. Savings in training time can be converted to savings in training costs to derive the direct economic benefit.

Reduced skill requirements. Improved work system structures and processes may also result in reducing the skill requirements required to perform some jobs. In addition to any savings that may result from reduced training requirements, employees with lower skill levels may be hired to perform the job, thus lowering the salary levels. The resultant salary savings constitute a direct economic benefit.

Reduced maintenance time. Improved work system structures and processes often result in reducing the system's maintenance requirements, requiring fewer maintenance personnel or enabling them to do other things. These savings in maintenance personnel can be converted into savings in salary and benefits to derive the economic benefit.

Reduced absenteeism. Reduction in lost time from employees who fail to show up to work for reasons other than accidents, injuries, or illness is another common outcome of effective macroergonomics interventions. Any savings in salaries and benefits for replacement personnel is a direct economic benefit. Reduced absenteeism also can result in (a) productivity increases, because there is less disruption of the work system and less work being done by replacement personnel (who are often less experienced and skilled in the specific job), and (b) reduced training, because fewer replacement personnel have to be trained.

Determining the Materials and Equipment Benefits

In addition to economic benefits related to increased employee productivity and reduced personnel expenses, macroergonomics improvements to work systems often result in materials and equipment savings. These include savings from reduced scrap, equipment, production parts and materials, and maintenance tools and materials.

Reduced scrap. Improved work system structures and processes can reduce production errors and the resultant production of defective items or wastage of materials. These savings can be calculated directly from the company's cost accounting data. This benefit may be shown on a per annum basis for a specified number of years.

Equipment savings. Improved work systems can also result in reducing the number of pieces of equipment required, performing a given function with less expensive equipment, or increasing the life of equipment because of better employee care and use. These cost savings can be calculated directly from the vendors' prices. In addition, there may be savings in reduced equipment installation and testing costs.

Reduced production parts and materials. Macroergonomics interventions sometimes mean that products can be produced with fewer parts or less expensive materials. These savings can be readily calculated based on the purchase price of the parts and materials or, if produced internally, the production cost.

Reduced maintenance tools and materials. Reductions in maintenance requirements can reduce not only personnel requirements but also the number of tools and amount of materials required for maintenance. The resulting economic benefit can be calculated in the same manner as used for production parts and materials, noted earlier.

The Structural Dimensions of Work Systems

We earlier noted that macroergonomics involves the development and application of human-organization interface technology and that this technology is concerned with improving the organizational structure and related processes of work systems. Accordingly, an understanding of macroergonomics requires an understanding of the key dimensions of organizational structure. As we show in Chapter 4, knowledge of the specific sociotechnical characteristics of a given work system will guide us in macroergonomically improving these key dimensions for that work system's organizational structure.

In order to provide a common framework, two basic concepts need to be clarified before we discuss the dimensions of a work system's organizational structure: *organization* and *organization design.*

Organization. An organization may be defined as "the planned coordination of two or more people who, functioning on a relatively continuous basis and through division of labor and a hierarchy of authority, seek to achieve a common goal or set of goals" (Robbins, 1983, p. 5). This definition can be broken down as follows:

1. Planned coordination of collective activities implies *management.*
2. Because organizations are made up of more than one person, individual activities must be designed and functionally allocated to be complementary, balanced, harmonized, and integrated to ensure an effectively functioning work *system.*
3. Organizations accomplish their activities and functions through division of labor and a hierarchy of authority. Thus, organizations have *structure.* How this work system structure is designed is critical to the organization's functioning.
4. The collective activities and functions of an organization are oriented toward achieving a common goal or set of goals. From a macroergonomics design standpoint, this implies that criteria for assessing an organization's design exist. They should be identified, weighted, and used in evaluating feasible alternative designs for the overall work system.

Organizational design. Organizational design refers to the design of an organization's work system *structure* and related *processes* to achieve the organization's goals. The organizational structure of a work system can be conceptualized as having three core dimensions: complexity, formalization, and centralization (Bedeian & Zammuto, 1991; Robbins, 1983; Stevenson, 1993). *Complexity* refers to the degree of differentiation and integration that exist within a work system. *Differentiation* refers to the extent to which the work system is segmented into parts. *Integration refers* to the number of mechanisms that exist to integrate the segmented parts for the purposes of communication, coordination, and control.

Complexity: Differentiation

Work system structures employ three common types of differentiation: *Vertical, horizontal,* and *spatial.* Increasing any one of these three increases a work system's complexity.

Vertical differentiation. Vertical differentiation is measured in terms of the number of hierarchical levels separating the chief executive position from the jobs directly involved with the system's output. In general, as the size of an organization increases, as measured by the number of employees, the need for greater vertical differentiation also increases (Mileti, Gillespie, & Haas, 1977). For example, in one study, size alone was found to account for 50%-59% of the vertical differentiation variance (Montanari, 1976).

A major reason for this strong relationship is the practical limitation of *span of control.* Any one manager is limited in the number of subordinates whom he or she can direct effectively (Robbins, 1983). Thus, as the number of first-level employees increases, the number of first-line supervisors also must increase. This, in turn, requires more supervisors at each successively higher level and ultimately results in the creation of more hierarchical levels in the work system's structure.

Although span-of-control limitations underlie the size-vertical differentiation relationship, it is important to note that these limitations can vary considerably, depending on a number of factors. So for an organization of a given size, if large spans of control are appropriate, the number of hierarchical levels will be fewer than if small spans of control are required. A major factor affecting span of control is the *degree of professionalism* (education and skill requirements) designed into employees' jobs. Generally, the higher the level of professionalism, the more employees are able to function autonomously and thus need less supervision. Consequently, the manager can effectively supervise a larger number of employees.

Other factors that affect span of control are the degree of formalization, type of technology, psychosocial variables, and environmental characteristics. These will be discussed separately in Chapter 4.

Horizontal differentiation. Horizontal differentiation refers to the degree of departmentalization and specialization within a work system. Horizontal differentiation increases complexity because it requires more sophisticated and expensive methods of control. In spite of this drawback, specialization is common to most work systems because of the inherent efficiencies in the division of labor. Adam Smith (1876/1970) demonstrated this point more than 200 years ago. He noted that 10 workers, each doing particular tasks (job specialization), could produce about 48,000 pins per day. But if each of the 10 worked separately and independently and performed all production tasks, they would be lucky to make 200 pins.

Division of labor creates groups of specialists, or *departmentalization.* The most common ways of designing departments into work systems are on the basis of (a) function, (b) simple numbers, (c) products or services, (d) client or client class served, (e) geography or spatial dispersion, and (f) process. Most large corporations use all six (Robbins, 1983).

Two of the most common ways to determine whether or not a work group should be divided into one or more departments are the degree of commonality of *goals* and of *time orientation.* Subgroups that differ either in goals or time orientations should be structured as separate departments. For example, sales department employees differ from research and development (R&D) employees on both dimensions; their output is different, and sales personnel usually operate on short time lines (one year or less), whereas R&D personnel operate on long ones (three or more years). Thus, they clearly should be departmentalized separately, and usually are (Robbins, 1983).

Spatial dispersion. Spatial dispersion refers to the degree to which an organization's activities are performed in multiple locations. There are three common measures of spatial dispersion: (a) the number of geographic locations that make up the total work system, (b) the average distance of the separated locations from the organization's headquarters, and (c) the proportion of employees in these separated units in relation to the number at headquarters (Hall, Haas, & Johnson, 1967). In general, complexity increases as any of these three measures increases.

Complexity: Integration

As noted earlier, *integration* refers to the number of mechanisms designed into a work system for ensuring communication, coordination, and control among the differentiated elements. As the differentiation of a work system increases, the need for integrating mechanisms also increases. This occurs because greater differentiation increases the number of units, levels, and so forth that must communicate with one another, coordinate their separate activities, and be controlled for efficient operation. Some of the more common integrating mechanisms that can be designed into a work system are formal rules and procedures, committees, task teams, liaison positions, and system integration offices. Computerized information and decision support systems can also be designed to serve as integrating mechanisms. Vertical differentiation in itself is a primary form of an integrating mechanism (i.e., a manager at one level serves to coordinate and control the activities of several lower-level groups).

Once the differentiation aspects of a work system's structure have been determined, a major task for the macroergonomics professional is to determine the kinds and number of integrating mechanisms to design into the work system. Too few integrating mechanisms will result in inadequate coordination and control among the differentiated elements; too many integrating mechanisms stifle efficient and effective work system functioning and usually increase costs. As we discuss in Chapter 2, a systematic analysis of the type of technology, personnel subsystem factors, and characteristics of the external environment can help to determine the optimal number and types of integrating mechanism.

Formalization

From an ergonomics design perspective, formalization can be defined as the degree to which jobs within the work system are standardized. Highly formalized designs allow for little employee discretion over what, when, or how work is to be accomplished (Robbins, 1983). In highly formalized designs, there are explicit job descriptions, extensive rules, and clearly defined procedures covering work processes. Ergonomists can increase formalization by designing jobs, machines, and software to standardize procedures and allow little opportunity for operator discretion. By the same token, human-job, human-machine, and human-software interfaces can be ergonomically designed to decrease formalization by permitting greater flexibility and scope to employee decision making (i.e., low formalization). When there is low formalization, employee behavior is relatively unprogrammed, and the work system allows for considerably greater use of one's mental abilities. Thus, greater reliance is placed on the employee's professionalism, and jobs tend to be more intrinsically motivating.

In general, the simpler and more repetitive the jobs to be designed into the work system, the higher should be the level of formalization. However, caution must be taken not to make the work system so highly formalized that jobs lack any intrinsic motivation, fail to effectively use employee skills, or degrade human dignity. Invariably, good macroergonomics design can avoid this extreme. The more nonroutine or unpredictable the work tasks and related decision making, the less amenable the work system is to high formalization. Accordingly, reliance has to be placed on designing into the organization a relatively high level of professionalism.

Centralization

Centralization refers to the degree to which formal decision making is concentrated in a relatively few individuals, groups, or levels, usually high in the organization. When the work system structure is highly centralized, lower-level supervisors and employees have only minimal input into the decisions affecting their jobs (Robbins, 1983). In highly decentralized work systems, decisions are delegated downward to the lowest level having the necessary expertise.

It is important to note that work systems carry out two basic forms of decision making: *strategic* and *tactical*. The degree of centralization is often different for each. Tactical decision making has to do with the day-to-day operation of the organization's business; strategic decision making concerns the long-range planning for the organization.

Under conditions of low formalization and high professionalism, tactical decision making may be highly decentralized, whereas strategic decision making may remain highly centralized. Under these conditions, it is also important to note that the information required for strategic decision making is often controlled and filtered by middle managers or even lower-level personnel. Thus, the more these people reduce, summarize, selectively omit, or embellish the information that gets fed to top management, the less is the actual degree of centralization of strategic decision making (Hendrick, 1997).

In general, centralization is desirable (a) when a comprehensive perspective is needed; (b) when it provides significant economies; (c) for financial, legal, or other decisions that clearly can be done more efficiently when centralized; (d) when operating in a highly stable and predictable external environment; and (e) when the decisions have little effect on employees' jobs or are of little interest to them.

Decentralized decision making is desirable (a) when an organization needs to respond rapidly to changing or unpredictable conditions; (b) when grassroots input to decisions is desirable; (c) to provide employees with greater intrinsic motivation, job satisfaction, and sense of self-worth; (d) when it can reduce stress and related health problems by giving employees greater control over their work; (e) to more fully utilize the mental capabilities and job-related knowledge of employees; (f) to gain greater employee commitment to, and support for, decisions by involving them in the process; (g) when it can avoid overtaxing a given manager's capacity for human information processing and decision making; and (h) to provide greater training opportunity for lower-level managers (Hendrick, 1997).

Chapter 2

MACROERGONOMICS THEORY

The Sociotechnical Systems Model

As noted earlier, macroergonomics is a *sociotechnical* systems approach to work system design. The sociotechnical systems model of work systems was empirically developed initially in the late 1940s and 1950s by Trist and Bamforth (1951) and their colleagues at the Tavistock Institute of Human Relations in the United Kingdom. Follow-on research by Katz and Kahn (e.g., 1966) of the Survey Research Center at the University of Michigan, and by many others, served to confirm and refine the sociotechnical systems model.

This model views organizations as transformative agencies; they transform inputs into outputs. Sociotechnical systems bring three elements to bear on this process: a *technological subsystem*, *personnel subsystem*, and *work system design* consisting of an organizational structure and processes. These three elements interact with one another and the external environment on which the organization depends for its survival and success. Insight into sociotechnical systems theory is provided by the classic Tavistock studies of Welsh deep-seam coal mining in the United Kingdom.

The Tavistock Studies

The formal origin of sociotechnical systems theory, and the coining of the term *sociotechnical systems*, can be traced back to the studies by Trist and Bamforth relative to the effects of technological change in a deep-seam Welsh coal mine (DeGreene, 1973). The traditional mining system relied largely on manual labor involving teams of small, autonomous groups of miners. Each group had control over its own work. Each miner performed a variety of tasks, and most jobs were interchangeable among workers. Considerable satisfaction was derived from being able to complete the entire task. In addition, through their close group interaction, workers could readily satisfy social needs on the job. Sociotechnically, the psychosocial and cultural characteristics of the workforce, the task requirements, and the work system's design were *congruent*.

The technological change involved replacing this more costly manual, or *shortwall*, method with mechanical coal cutters. No longer restricted to working a short face of coal, miners could extract coal from a long wall.

However, this new and more technologically efficient longwall system resulted in numerous changes. The work system design was no longer congruent with workers' psychosocial and cultural characteristics: Miners no longer worked in small groups; instead, shifts of 10 to 20 men were required. Jobs were designed to include a set of narrowly defined tasks. Opportunities for social interaction were severely limited, and job rotation was not possible. There was a high degree of interdependence among the tasks of the three shifts, and problems from one shift carried over to the next, thereby holding up labor stages in the extraction process. This complex and rigid work system design was very sensitive to productivity and social disruptions. Instead of the expected improved productivity, low production, absenteeism, and intergroup rivalry became common (DeGreene, 1973).

In follow-on studies of other coal mines by the Tavistock Institute (Trist, Higgin, Murray, & Pollock, 1963), this conventional longwall method was compared with a *composite* longwall method. In the composite method, the work system design utilized a combination of the new technology and features of the old psychosocial work structure of the manual system. The composite longwall system reduced interdependence among the shifts, increased the variety of skills used by each worker, created opportunities for satisfying social needs on the job, and permitted teams to select their own members. Production was significantly higher than for either the conventional longwall or the old manual system. Absenteeism and other measures of poor morale and dissatisfaction dropped dramatically (DeGreene, 1973).

Based on the Tavistock Institute studies, Emery and Trist (1960) concluded that *different work system designs can utilize the same technology*. The key is to select a work system design that is compatible with the characteristics of the people who will perform tasks and the relevant external environment, and then employ the available technology in a manner that achieves congruence with it.

From our experiences, although any given technology can be used with different work system designs, technology, *once it is employed in the system*, does constrain the subset of possible work system designs. Computers and related automation in managerial, administrative, production, logistical, marketing, and other facets of modern complex systems constrain the range of possible work system designs more than in more traditional labor-intensive work systems. It thus has become increasingly important first to determine the optimal *macroergonomic* design of the work system before fully proceeding with the micro-ergonomic design of human-machine/software modules, subsystems, and interfaces.

For example, Hendrick consulted with a telecommunications company that had developed a complex software program for conducting a major operational process. The software procedure prescribed a type of work system design that was not compatible with the organizational unit's sociotechnical characteristics. As a result, less than two years later, the software system was scrapped, an appropriate work system design was determined, and *then* a new software program was designed that was compatible with the work system. In short, at least conceptually, "a top-down ergonomic approach is essential to ensure that the dog wags the tail, and not vice versa" (Hendrick, 1984).

Joint Causation and Subsystem Optimization

The sociotechnical system concept views organizations as open systems engaged in transforming inputs into desired outputs (DeGreene, 1973). *Open* means that work systems have permeable boundaries exposed to the environments in which they exist (political, economic, social, etc.). These environments thus enter or permeate the organization along with the inputs to be transformed. The primary ways in which external environmental changes enter the organization are through (a) its marketing or sales function, (b) the people who work in it, and (c) its materials or other input functions (Davis, 1982).

As transformative agencies, organizations continually interact with their external environment, receiving inputs, transforming them into desired outputs, and exporting these outputs to their environment. They are thus both influenced by and influence the external environment.

In performing this transformative process, organizations bring two critical factors to bear on the transformation process: technology in the form of a *technological subsystem*, and people in the form of a *personnel subsystem*. In general, the design of the technological subsystem primarily defines the *tasks* to be performed, whereas the design of the personnel subsystem prescribes the *ways* in which tasks are performed. Each interacts with the other at every human-machine and human-software interface. The technological and personnel subsystems thus are mutually interdependent. Both subsystems operate under *joint causation*, meaning that they are affected by causal events in the external environment (such as consumer preferences, competition, and changes in government regulations).

Joint causation leads to the related sociotechnical system concept of *joint optimization* of the work system in terms of its ability to accomplish the transformative process. Of importance to macroergonomics is the fact that the technological subsystem, once designed, is relatively stable and fixed. It thus falls to the personnel subsystem to adapt further to environmental change.

Because the technological and personnel subsystems respond jointly to causal events, designing one subsystem and then fitting the second to it usually results in a joint work system that cannot respond effectively to its external environment (i.e., it is suboptimal). Maximizing overall work system effectiveness requires the *joint design* of the technical and personnel subsystems in order to develop the best possible fit between the two, given the objectives and requirements of each subsystem and of the overall work system (Davis, 1982). Inherent in this joint design is developing an optimal structure for the overall work system so it will be compatible with the organization's sociotechnical characteristics.

Joint Optimization versus Human-Centered Interface Design

At first glance, the concept of joint optimization may appear to be at odds with human-centered interface design. In human factors/ergonomics, we traditionally speak of changing the person through selection or training, or changing the system through design, but in both paradigms we are designing to support the human. Does jointly optimizing the technological subsystem compromise the human's dominant position in the human-machine relationship? Perhaps, as envisioned by the original nonergonomist researchers and practitioners in the post World War II sociotechnical systems movement, the social system (including the human) had equal billing with the technical system – at least in theory. In practice, though, we have heard the sociotechnical systems movement criticized for being technology driven.

In the macroergonomics subdiscipline, the sociotechnical systems literature provides a theoretical foundation. But as ergonomists, we emphatically maintain that ergonomists design to support human capabilities and limitations and studiously avoid a technology-driven design mindset. For ergonomists, the principle of joint optimization is the avoidance of optimizing – or, more precisely, maximizing – any single sociotechnical system element. Maximizing the *technological* subsystem, as when humans are given the "leftover" functions to perform in an automated system, suboptimizes the overall system. Maximizing the *personnel* subsystem, as when expensive behavior and attitude modification training is undertaken without consideration of the organization's technology or other sociotechnical characteristics, suboptimizes the overall system. Attempting to maximize the *organizational* design by constantly restructuring without macroergonomically sound cause (e.g., for purely political reasons, such as creating a new subunit to reward a friend of the boss with a managerial position) suboptimizes the overall system. Finally, maximizing the *external environment* by allotting too much time with external stakeholders (e.g., customers, suppliers, stockholders, government agencies) at the expense of internal operations will suboptimize the total system.

To achieve the appropriate balance, then, joint optimization is achieved through (a) joint design, (b) a human-centered approach to function and task allocation/design, and (c) attending to the organization's sociotechnical characteristics. Chapter 3 offers some pragmatic methods to achieve jointly optimized sociotechnical work systems.

The Sociotechnical System Elements

Each of these components (technological subsystem, personnel subsystem, and relevant external environments) has been studied in relation to its effects on the three organizational design dimensions described in Chapter 1 (complexity, formalization, and centralization), and empirical models have emerged. These models can be used as macroergonomics tools in analyzing organizations and developing or modifying their work system designs. Some of these models have proven particularly useful in macroergonomics analyses and interventions and will be described in Chapter 4.

A consistent finding, and one that is fundamental to sociotechnical theory, is that the four basic sociotechnical system elements (technological subsystem, personnel subsystem, external environment, and work system design) are mutually interdependent. If some characteristic of one of the four elements is changed, it will affect the other three. Thus, if some aspect of the personnel subsystem is changed, it will affect the technological subsystem, the work system's interaction with the external environment, and the structure and/or processes of the work system. If these impacts on the other three sociotechnical system elements are not anticipated and planned for, the result is likely to be a dysfunctional or suboptimal work system. This is critically important.

Based on assessments of more than 200 organizational units, Hendrick has observed that managers generally fail to recognize the interdependence of the four sociotechnical system elements. All too often, a manager sees a problem with some aspect of one of the sociotechnical system elements and then attempts to fix that specific work system problem. The organization then experiences a series of ripple effects on the other sociotechnical system elements that, in turn, create other problems. Not infrequently, these other problems are greater than the one the manager attempted to fix. In one such case in a hazardous industry, as a direct result of the senior manager's "fix" (a personnel policy change to meet an arbitrary production goal that caused operators to work overtime even when fatigued), the serious accident rate went from an average of one per year to over one per month!

Relation of Macro- to Micro-Ergonomics Design

Once a systematic macroergonomics approach has been used to determine the design characteristics of the overall work system, the next step is to carry through that design to the micro-ergonomic level. The act of defining the design characteristics of the overall work system prescribes many of the characteristics of the jobs to be designed into the system and of the related human-machine and human software interfaces. Some examples follow (Hendrick, 1991).

1. As noted in Chapter 1, decisions about an organization's horizontal differentiation (the degree of departmentalization and specialization within a work system) determine how narrowly or broadly jobs are defined and how they should be grouped or departmentalized.

2. Decisions concerning the level of formalization and centralization will dictate (a) the degree of routinization and employee discretion to be ergonomically designed into the jobs and attendant human-machine and human-software interfaces, and, consequently, (b) the level of professionalism to be designed into each job. These decisions also help prescribe many of the design requirements for the information, communications, and decision support systems, including what kinds of information are required by each job and networking requirements among them.

3. Decisions about the organization's vertical differentiation (the number of hierarchical levels separating the chief executive position from the jobs directly involved with the system's output) largely determine the number of managers required by the work system. When these decisions are coupled with those concerning horizontal differentiation, spatial dispersion, centralization, and formalization (see Chapter 1), they collectively prescribe many of the design characteristics of the managerial positions. These characteristics include span of control, decision authority, nature of the decisions to be made, information and decision support requirements, and qualitative and quantitative educational and experience requirements.

In summary, effective macroergonomic design drives a number of aspects of the micro-ergonomic design of the work system and thus ensures *ergonomic compatibility* of the system components with the work system's overall structure. In sociotechnical system terms, this approach enables joint optimization of the technical and personnel subsystems from top to bottom throughout the organization and harmonization of the work system's elements with its overall design and the external environment. The result is greater assurance of optimal *system* functioning and effectiveness, including productivity, quality, and employee safety and health, psychosocial comfort, intrinsic motivation, commitment, and perceived quality of work life.

In contrast, either a purely micro-ergonomic approach or, more commonly, a technology-centered approach is very likely to create work systems in which the personnel subsystem is forced to adapt to the system's technology and structure in a manner similar to trying to fit a square peg into a round hole. Beginning with the longwall coal mining studies by the Tavistock Institute (Trist & Bamforth, 1951), the organizational literature is full of examples that consistently show how this lack of compatibility adversely affects system productivity and, along with it, employee job satisfaction and commitment (e.g., see Argyris, 1971, for a summary of the findings of a number of prominent U.S. researchers).

Because the organization, jobs, and human-machine and human-software interfaces often already exist, once the design modifications of the overall work system are determined, the next step is to review those jobs and interfaces to determine if they are congruent with the modified work system. Where they are not, decisions have to be made about how to modify them to make them congruent. As we note later, this aspect of the macroergonomics process should involve active employee participation.

Systems Theory and Organizational Synergism

A widely accepted view among system theorists and researchers is that all complex systems are *synergistic*: The whole is not equal to the simple sum of its parts. Because organizations are complex systems, they too should be synergistic. Theoretically, because of this synergism, certain circumstances should tend to occur in complex work systems (Hendrick, 1994), described in the following sections.

When Work Systems Have Incompatible Designs

When work system structures and processes are grossly incompatible with their sociotechnical system characteristics, and/or jobs and human-system interfaces are incompatible with the organization's structure, the whole is less than the sum of its parts. Under these conditions, we can expect (a) productivity – especially *quality* of production – to be relatively deficient, (b) accident rates and lost-time injuries to be relatively high and adherence to safety standards and procedures poor, and (c) motivation and related aspects of job satisfaction and perceived quality of work life (e.g., psychosocial comfort, stress) to be relatively poor.

When Work Systems Have Compatible Designs

When a work system has been designed effectively from a macroergonomics perspective, and that effort has been carried through to the micro-ergonomic design of jobs and human-machine and human-software interfaces, then the work system design is *harmonized*. As a result, synergistic functioning becomes possible, and the various system effectiveness criteria, such as productivity, safety, employee satisfaction, commitment, and perceived quality of work life, will be much greater than the simple sum of the parts.

Implications for the Potential of Organizations

Assuming that these two theoretical propositions are true, then macroergonomics has the potential to greatly improve productivity, safety, health, employee motivation and commitment, and the quality of work life. In the early 1990s, Hendrick theorized that instead of the 10%–25% improvements in these system effectiveness measures that many ergonomists have experienced from successful micro-ergonomics interventions, one should see improvements of 60%–90%, or more (Hendrick, 1994). As we document with actual cases in Chapter 6, this prediction is proving to be accurate.

Chapter 3

COMMON MACROERGONOMICS METHODS

Macroergonomics methods are continuing to be developed, used, and vali-dated. In this chapter, we scratch the surface of the plethora of methods cur-rently in use to optimize work system design. First, we introduce *participatory ergonomics* as a staple approach in macroergonomics analysis and design. Then we briefly review other methods used in macroergonomics.

Participatory Ergonomics

As previously characterized, macroergonomics is top-down (i.e., strategic approach to analysis), bottom-up (e.g., participatory) and middle-out (i.e., focus on processes). Central to macroergonomics is the expectation that analysis and design of work systems will be participatory in nature (Imada, Noro, & Nagamachi, 1986). To achieve human-centered work system de-signs, human-centered analytical processes must be used. This constitutes the sociotechnical principle of *compatibility*. Consistent with the emphasis on participation in a number of domains, participatory ergonomics is rapidly emerging as an area of international inquiry in its own right (Brown, 1994) and is seen as a core method in macroergonomics.

Some have suggested that a broader context for participatory ergonomics is the *participative management* movement (e.g., Wilson & Haines, 1997) or the worker participation movement (Cohen, 1996). Historically, the notion of participation probably has its roots in *human relations theory* (circa 1930s), which acknowledged the importance of workers' feelings and attitudes in work. Certainly, decentralization of decision making was also an important tenet in sociotechnical systems design in the 1950s, as evidenced by the devel-opment of autonomous work teams, the precursor to today's "self-managed" or "high-performance" work teams.

In addition, participation was a core construct in the quality of work life movement of the 1970s (e.g., Cherns & Davis, 1975). Today's participation can take the structural form of a nominal group, a cohesive team, or individual involvement. Indeed, Japanese quality circles, total quality management (TQM) process improvement teams, and the like imply some decentralization of the decision-making environment. Kleiner (1999) illustrated how macroergonomics can be an organizing framework for ergonomics and TQM, so macro-ergonomics and TQM are not necessarily mutually exclusive.

For the purposes of this discussion, when the participation or involvement involves ergonomic analysis or design, the particular type of employee involvement can be said to constitute participatory ergonomics. Participatory ergonomics is defined as follows:

> the involvement of people in planning and controlling a significant amount of their own work activities, with sufficient knowledge and power to influence both processes and outcomes in order to achieve desirable goals. (Wilson, 1995, p. 37)

Although at first glance this definition is fairly straightforward, participatory ergonomics is easier said than done, can take a variety of forms, and can take different forms in different cultures (Haines & Wilson, 1998). This definition implies an organizational structure to support ergonomics, whereby workers are involved in planning; control their own work activities through, for example, job design; have sufficient education, training, and development; and are empowered or have sufficient decision-making authority.

The following sections briefly present some of the complexities associated with the structure, process, requirements, applications, and potential costs and benefits of participatory ergonomics.

The Structure and Process of Participatory Ergonomics

Figure 3.1 illustrates a general framework for structuring participatory ergonomics initiatives (Haines & Wilson, 1998). First, an organizational decision is made to implement some form of participatory ergonomics. According to Haines and Wilson (1998), several motivational factors contribute to this decision, including the availability of expert advice, external recommendations, workforce/union negotiations, management philosophy, awareness of ergonomics problems, complaints and claims, and product/market competitive advantage. Following the decision to implement participatory ergonomics, there is a need to structure the initiative. Factors that influence the design include organizational size and culture, nature of the ergonomics problems, time frame available, stakeholders, available resources, and training/education.

Once a decision to implement has been reached, implementation methods can be chosen. These will be a function of type of problem, resources available, knowledge of/experience with methods, and number of participants. Finally, a feedback loop is required for ongoing *continuous improvement*, the sociotechnical principle of *incompletion*. In the macroergonomics perspective, this is an "open" system, subject to factors from the outside environment

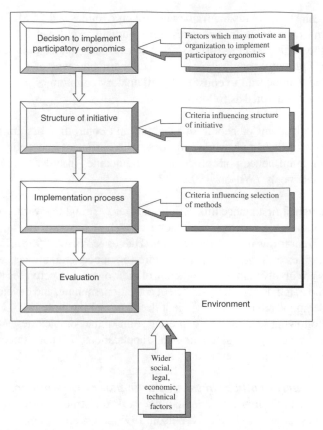

Figure 3.1 General framework for developing and implementing a participatory ergonomics initiative (from Haines & Wilson, 1998).

(e.g., social, economic, legal, technical). Unlike other methods that might acknowledge the existence of these factors, during macroergonomics design, the factors are analyzed and considered.

Participatory ergonomics can vary according to the level at which it is applied. At the organizational level, ergonomics programs are often sustained by integration with existing programs and structures. Within this context, operators can participate in the analysis, redesign, and evaluation of their jobs, tasks, and workplaces (Wilson, 1995). Organizationally, in our experience, ergonomics programs typically are administered by such departments as health and safety, operations, engineering, or human resources. Participatory ergonomics originally took shape within the context of quality of life programs (Brown, 1986), but in many organizations today, participatory ergonomics has been integrated with total quality management and other work process improvement efforts. Therefore, the macroergonomics approach recognizes that the unit of work is often the group or team and not the individual; in addition, it recognizes that the context for change may be other organizational change programs. Thus, the macroergonomist may become a work system integrator.

Participation can take many structural forms: quality circles, labor-management committees, cross-functional teams, self-managed teams, and individual participation. *Quality circles* have been used effectively to combat quality problems on the shop floor. These permanent problem-solving teams solicit the involvement of operators in solving quality issues. *Labor-management committees* address many issues, including health and safety, quality management, and strategic planning. *Cross-functional teams* are characteristic of total quality management environments in which workers and professionals from different functional organizations join forces to improve cross-functional technical and business processes. Cross-functional teams are parallel structures; that is, they coexist with the formal organizational structure. *Self-managed or autonomous teams* are used in high-performance work systems in which decentralization is high. Unlike parallel cross-functional teams, they are formal in nature (i.e., they are part of the recognized organization chart). *Individual participation* has been used successfully in ergonomics problem solving, but participation has been accomplished using groups as well.

Participatory ergonomics can also vary with respect to purpose, continuity, involvement, formality, requirement, decision making, and *coupling* (Haines & Wilson, 1998):

Purpose: An important distinction is whether participatory ergonomics is to be used to implement a particular change or whether it is the method of work organization itself (Wilson & Haines, 1997) – that is, whether it is the means to the end or the end itself.

Continuity: Continuity is dependent on whether the participatory ergonomics process is to be used in a continuous (daily) or discrete (periodic) timeline (Wilson & Haines, 1997).

Involvement: This characteristic involves whether participation is direct or representative (Haines and Wilson,1998). With *full direct participation*, all those affected by a decision or design become involved. *Partial direct participation* occurs when a representative subset of participants is involved, which is usually necessary because of economic considerations. Focus groups are an example of representative involvement.

Formality: Formality is determined by whether and to what extent involvement is formal, as in the case of permanent self-managed teams, or informal, in the case of a temporary, parallel team, or when the manager solicits input from workers on an ad hoc basis.

Requirement: Related to formality, involvement is either required or voluntary. A measure of this characteristic is whether workers are evaluated and rewarded on the basis of their participation. A related work design issue that typically arises is *when* the involvement will occur. That is, will workers participate as part of their formal jobs, or will they be expected to participate on an overtime basis? If the latter, will they be compensated for their extracurricular efforts?

Decision making: Decision-making differences relate to the extent of decentralization or empowerment available.

Coupling: Coupling refers to whether participative methods are directly or remotely coupled. In the former, there is little or no filtering of participant input; in the latter there is some filtering or translation, usually by managers or consultants (Haines & Wilson, 1998; Wilson & Haines, 1997).

Applications of Participation

Participation in decision making and problem solving. Nickerson (1992) identified decision making as an important application area for the ergonomist. At the organizational level, decision making takes the form of planning. As reported by Kleiner (1997), formalized, detailed planning has been criticized. One of the concerns posed by Mintzberg (1994) was that those who must implement plans are seldom involved in the decision-making processes inherent in the development of such plans. Not only will commitment be greater if implementers are involved, but more valid decisions will likely be reached because of the implementers' better knowledge of operational conditions, constraints, and opportunities.

Regarding operational and tactical decisions, it is assumed that operators have some level of decision-making responsibility in team-based organizational environments. Functional and cross-functional teams have, at most, the authority to make recommendations; self-managed teams have final decision authority. According to Sullivan (1999), 84% of top executives believe employees have adopted a better team mind-set in the past five years, and 79% of top executives claim self-managed teams will increase productivity for U.S. companies. Although the trend is toward self-managing teams, the total percentage of teams in this category is still fairly low. It is claimed that 10 years of improvement can be made within a single year with such teams (Carroll, 2000)!

Supporting the "empowered" human decision maker or decision-making team with appropriate tools and techniques is another task for the macroergonomist. For example, many computer-based tools are available to promote collaboration and participation in decision making. Cano, Meredith, and Kleiner (1998) categorized these as *presence support, communication support,* and *decision support technologies.* Presence support is the goal of such technologies as videoconferencing, in which distributed designers need to participate on a common design, or a remote individual participates in a design project. Communication support technologies actually facilitate collaboration, group dynamics, and the ability to transfer information. Communication and decision support technologies actually aid in the decision-making process by converting data to information and decision making.

Participation in product and system design. Participatory design has emerged as a popular approach in the area of usability evaluation of both products (including software) and systems. Participation of users in software and interface design is now a recognized method in the *user-centered design* life cycle. Snow, Kies, and Williges (1997) used the approach to design a physical environment (a new laboratory), utilizing virtual walkthroughs for participation. Sociotechnical approaches to information system design and implementation have also specified that users participate (e.g., Eason, 1988). Such involvement has led to jointly optimized work systems in which neither the personnel nor technological subsystems dominate, resulting in more positive organizational outcomes.

Participation in training design. In the paradigm of ergonomics, to improve performance, either the human can be selected or trained or the system can be redesigned. The preferred approach is to design systems to support human capabilities and limitations. Ideally, education, training, and development are then used to support the human in an ergonomically designed system, as opposed to helping the human adapt to a poorly designed or inflexible system. As work increasingly becomes knowledge-dependent (in part because of the growth of information technology), knowledge-based education as well as skills-based training will become increasingly important (Drucker, 1995).

Involving representative trainees in the design and development of training programs is a form of participation that should be considered by the macroergonomist. Additionally, training programs themselves should be designed with flexibility to support trainees' needs. Ideally, trainees can proceed through a given training experience on a participatory basis, selecting modules that relate to their particular needs and providing valuable evaluation feedback to training system designers.

Participation in work system analysis and design. Job involvement focuses on the design of work in order to produce better performance and quality of work life (Brown, 1996). *Job enrichment* and *job enlargement* are traditional ways to modify job design (e.g., see Organ & Bateman, 1991; Szilagyi & Wallace, 1990). Although ergonomics support is most widely used in workstation design, ergonomists have played a major role in organizing team-based work systems, plant layout, inspection methods, statistical process control, material handling, and work methods (that is, the involvement in and integration of broader system components).

Worker participation in work redesign has been shown to be desirable for enhancing the quality of working life (Baitsch & Frei, 1984). However, macroergonomics goes beyond traditional job design and ergonomics to achieve a higher level of involvement (e.g., Brown, 1996), a level at which the worker helps to analyze and design at the individual, group, and organizational levels.

One methodology to achieve this high involvement is discussed in Chapter 5: *analysis and design of work system process.* In this methodology, stakeholders participate in the environmental analysis and system analysis activities. This expands previous sociotechnical methods of analysis (e.g., Emery & Trist, 1978) to emphasize the role of participation, especially at system, function, and task levels. Others within the work system participate in every step of a 10-step process. This includes developing an understanding of their own roles and the roles of those with whom they interact; the tools, information, and interfaces needed for effective interaction; and the requisite skills and knowledge required to perform the job or redesign. This approach recognizes that participation is critical in function allocation, task allocation, and other system design decisions (Clegg, Ravden, Corbertt, & Johnson, 1989).

Participatory Ergonomics Requirements

Before discussing the benefits and costs of participatory ergonomics, it is worth repeating that participatory ergonomics is easier said than done. There are several critical requirements, which, if left unattended, will compromise the results. First, it is imperative that senior managers and supervisory personnel support any participatory program in the work system. Although participatory ergonomics can operate at several levels of the organization, with several possible objectives (Wilson & Haines, 1997), organizational support is still important. Ideally, a *culture* of participation should exist, in which a clear organizational guiding principle, expressed and practiced, is valuing employees' involvement. Sufficient organizational support is needed (Wilson & Haines, 1997), and an appropriate facilitator is also recommended (Wilson & Haines, 1997).

Appropriate processes and methods, continuous improvement, and institutionalization are all components of a successful participatory ergonomics program (Wilson & Haines, 1997). Depending on the extent of the participation, requirements can vary in degree. For example, in the *high-involvement* approach, workers are not only involved in their own job or group design but also have a role in organizational design and performance (Brown, 1996). To achieve an organizational commitment, workers will need exposure to additional knowledge and information about the overall organization, especially if they are involved in decisions about strategy and structure (Brown, 1996).

Along with training, then, the necessary information is required in a participatory environment. In our experience, it is also important that the worker or group share management's mental model about level of authority, and specifically whether the participation will result in recommendations or actual decisions to be implemented. As Van Aken and Kleiner (1997) indicated in their study of design teams, getting the right mix of skills in a group or team is a very important determinant of success. Also, feedback/evaluation mechanisms need to be in place to continuously improve the participative efforts. Feedback is instrumental to teams as they adjust their shared mental models of performance. Over time, there may be a need to adjust the participatory ergonomics cycle of involvement (Haines & Wilson, 1998), based on the evaluation feedback, by changing the skill mix, group process, and so on. Participants' motivation, confidence, and competence are all areas that may require intervention, as illustrated in Figure 3.2.

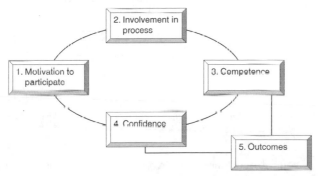

Figure 3.2 Participatory ergonomics cycle (adapted from Haines & Wilson, 1998).

Finally, we are compelled to mention the emerging issues related to distributed or geographically dispersed participation. If the participation is achieved by connecting distributed workers via information and communications technology, then additional complexities will require attention. A sense of "presence" will need to be achieved in order for the participant to feel truly involved (Cano, Meredith, & Kleiner, 1998). This sense of presence can be accomplished by using the correct technologies at the correct times in the correct ways. For example, in resolving a conflict, the optimal form of presence would be established through face-to-face communication. If this is not possible, high-fidelity audio and video must be used to create reliable visual and auditory cues needed to develop a shared understanding of the communication. Conversely, text-based communication, such as electronic mail, is an inappropriate and ineffective medium for ensuring a shared experience or presence in conflict resolution. In the end, there is still no substitution for building periodic face-to-face meetings into the process.

Participatory Ergonomics Benefits and Costs

There is little doubt that the participative nature of macroergonomics intervention accounts for a good deal of the impressive results achieved (see Chapter 6 case studies). For example, Haines and Wilson (1998) theorized that participatory ergonomics can play a substantive role in health and safety, as illustrated in Figure 3.3.

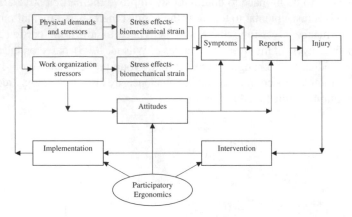

Figure 3.3 Potential role of participatory ergonomics in health and safety (Haines & Wilson, 1998).

Having control over one's own work has its motivational benefits. To achieve the benefits of work quality and commitment, several conditions must exist (Cohen, 1996): The task must be viewed as important and not trivial, the work must be interesting and somewhat challenging, and workers must be properly trained for the participative work.

For more than routine decisions, it is fairly well established that participation leads to higher-quality decisions. Beyond the traditional benefits of commitment and quality, a richer learning experience for designers, planners, users, and workers may result (Wilson & Haines, 1997). When the type of decision making is a design activity, for example, more alternatives and ideas tend to be generated. In some environments, such as *agile* or *flexible manufacturing*, participation is directly related to the flexibility needed in production. In addition, having more participants allows for more information to be accessed in problem solving and decision making. Finally, participation adds value by providing integration in organizational design. Thus, communication and coordination are enhanced as a result of participatory processes.

Participatory ergonomics shares some of the potential pitfalls found in other forms of organizational involvement. One of the most unfortunate situations occurs when an individual or team misperceives its actual level of authority. Level of authority has many degrees, but for illustration purposes, we envision three discrete levels of decentralized authority: none (autocracy), some (participatory recommendation), and complete (delegation). No participation is appropriate when decisions are either so mundane or so time-critical that it is desirable to have a single person (e.g., the supervisor) make the decision. This is partly because participation takes time. When quality and/ or commitment are important (as in the case of ergonomics or work system design), some degree of participation is appropriate. Participation is also desirable when problem, decision, or design complexity requires a multitude of skills and knowledge sets. When supervisors can live with the decisions of others (as in the case of autonomous groups), complete delegation may be appropriate.

Problems arise when expectations about level of authority are not clear. For example, if an individual or group is asked to participate in an ergonomics design process, the natural assumption will be that implementation will be a certainty, because workers tend to assume they have more authority than has been delegated. However, if managers really intended the participation to culminate in a design recommendation for possible consideration, great disappointment will follow a management decision not to implement. This miscommunication creates additional barriers for future involvement.

Certainly, there are times when participation is desired for the sole purpose of "buy-in" without any real attention to workers' input. Wilson and Haines (1997) called this "cynical" or "half-hearted" application. Any short-term gains of commitment will surely erode in such an environment, given sufficient time. Those doing quality of working life research in the 1970s had the perspective that the relative importance of quality of working life to productivity was an important determinant of long-term success, especially if inherent trade-offs existed between the two goals (e.g., Herbst, 1975).

Other potential problems with participation include individual or group resistance, unwillingness or inability to participate, and the nature and structure of the participative efforts themselves (Wilson & Haines, 1997).

The whole notion of participation can be elusive. We summarize this section by emphasizing that at its core, participation is a *value*. Assuming managers believe that users, workers, trainees, and others should participate in decisions that affect them and have the skill and knowledge to contribute, then particular participatory macroergonomics methods, tools, and techniques are available to solicit this involvement in analysis and design methods. A discussion of additional methods follows; many of these assume participation by individuals or groups.

Other Common Macroergonomics Methods

A number of methods frequently used in various kinds of organizational studies have been adopted for use in macroergonomics analysis, intervention, and evaluation. These include the *laboratory experiment, field study, field experiment, organizational questionnaire survey, interview survey,* and *focus group* (Hendrick, 1997).

Laboratory Experiment

The laboratory experiment – or laboratory study, as it is variously called – is the scientist's primary approach for determining causal relationships. The basic approach is to manipulate some independent variable while controlling or otherwise discounting the effects of other variables that might affect the dependent variable. Thus, when changes occur in the dependent variable, the cause of the change can more clearly and efficiently be attributed to the independent variable than to other research approaches, such as those described later.

Advantages. The major advantages of the laboratory experiment are that the researcher can (a) be fully prepared for accurate observations, (b) repeat the observations under the same conditions, (c) enable other experimenters to duplicate the results and make an independent check of them, and (d) vary the conditions systematically and note the variation in results.

Disadvantages. The primary disadvantage of the laboratory approach is its lack of realism. In real life, things do not occur in isolation. In fact, often there are *interactions* among the causal variables that determine the effects on the dependent or outcome variables. Bringing things into the sterile environment of the laboratory thus may change their nature – not only of the situation but also of the *motivation* of the participants, which may be different from those in the field environment. Typically, laboratory participants may not be representative of the actual worker populations they are intended to represent (e.g., college sophomores majoring in psychology may be representative of undergraduate psychology majors, but not industrial or office worker populations).

Macroergonomics applications. In reality, the magnitude of the realism shortcoming depends, to some extent, on the nature of the problem and one's ability to simulate field conditions in the lab. In macroergonomics applications, we usually attempt to create a simulated work system environment in the laboratory that enables us to systematically manipulate either work system or sociotechnical variables of interest (e.g., complexity, centralization, formalization), and then systematically observe and record the impact on various performance variables of interest (see Chapter 6).

Field Study Method

The field study approach is variously referred to as *systematic or naturalistic observation* and as *real-life research*. All these terms, when taken together, provide a good description of this approach. It involves going out into the field to systematically observe events as they occur naturally in real life.

Advantages. The primary advantage of the field study method is realism. By observing things as they occur naturally, the scientist avoids the sterility and artificiality of the laboratory. However, it is important for the researcher to recognize that his or her very presence changes the situation and thus can effect what happens. Great care has to be taken to be as unobtrusive as possible.

A second advantage is that when one is able to establish cause-effect relationships in the field setting, one can have high confidence in the practical usefulness of the results.

Disadvantages. The primary disadvantage is that the researcher has to wait for things to occur naturally. Thus, the observation process may take a long time and incur considerable expense before the cause-effect relationships can be established. A related disadvantage is that one may have to observe things occurring naturally many times, and under different conditions, before extraneous variables can be eliminated as causal factors and the true causal variables and their interactions can be identified.

Macroergonomics applications. By analyzing existing performance records for a given work system, and also by studying the characteristics of that work system and the related sociotechnical characteristics of the organization (e.g., see Chapter 4), we can usually identify problem areas that are amenable to macroergonomics interventions and gain insight into the nature of the macroergonomics intervention that is needed (see Chapter 6).

Field Experiment Method

One of the most widely used methods in macroergonomics interventions is the field experiment. Unlike the field study, the field experiment does not merely observe events as they occur naturally. Rather, the researcher or change agent deliberately and systematically manipulates variables and then observes the effect on the outcome or performance variable(s) of interest.

Advantages. In addition to the field study's advantage of realism, the field experiment also overcomes a major disadvantage of the field study: You do not have to wait for things to happen. Thus, in comparison with the field study, the field experiment is more efficient in terms of time and related costs. Further, the researcher or change agent can effect the specific changes of interest, which might never occur naturally.

Disadvantages. By causing a desired change to occur, the researcher or change agent may be introducing extraneous variables that influence the effects of the change. For example, in macroergonomics field experiments, employees' perceptions of the purpose of the change can alter how they respond to it and their related motivation. Also, *how* changes are implemented often determines the success or failure of the intervention. Finally, organizations could find the cost of using workers in experiments or quasi-experiments to be prohibitive.

Macroergonomics applications. The field experiment also can be used as a follow-on to an initial field study as a macroergonomics intervention. Sometimes this will be done in a particular part of the organization to test market it and then, if it proves effective, implement it on a larger scale. In part, because of the potential problems of employee acceptance and support of the intervention, the field experiment often is combined with a participatory ergonomics approach to effect the intervention (i.e., employees are far more likely to accept and support changes when they are actively involved in the change process).

Organizational Questionnaire Surveys

Perhaps the most widely used method for evaluating organizational functioning and identifying deficiencies is the written survey questionnaire. Some form of periodic questionnaire is used by most large corporations and many smaller ones. A number of generic survey questionnaires have been developed and validated for assessing such things as organizational climate, supervisory or managerial behavior and practices, and attitudes about such things as safety, fringe benefits, and one's job. In addition, organizations will develop their own surveys, or supplemental questions, to assess specific issues of interest, such as attitudes about a recent or proposed organizational change or about incentive systems.

A widely used variation of the organizational questionnaire survey method is the *survey feedback* method. With this method, data gained from the survey are summarized and subgrouped statistically and by organizational level, department, or project and then are "fed back" to the individual organizational units. Most commonly, the unit manager then participates with the unit's employees in interpreting the data and deciding on what changes, if any, should be made to improve unit effectiveness.

Constructing an organizational survey that can yield valid, useful data is a very tricky process and requires the help of a trained expert, such as an industrial/organizational psychologist. Even with expert help in constructing the questionnaire, it is essential to pilot-test the survey with a small, representative group to make sure both the instructions and questions are understood as intended. Invariably, refinements will be required as a result of the pilot test.

Advantages. A major advantage of this method of gathering information is anonymity and confidentiality, particularly for employees. If the survey is designed to protect anonymity and confidentiality, employees are likely to feel freer to express their opinions without fear of reprisal, and important information that is not part of the existing management information system can be obtained.

A second major advantage is that the opinions and perspectives of a large number of employees can be obtained relatively quickly and inexpensively. Thus, it is easy to ensure a large representative sampling of the employees, or to obtain the results for the entire workforce.

A third advantage is *replication*: The same questions can be asked over time and changes in response measured.

Finally, questionnaire data lend themselves to *quantitative analysis.* This enables comparison of the data for various work system units, levels, or subgroups and makes it possible to note areas of both consensus and variation.

Disadvantages. One major potential disadvantage is ambiguity of purpose. Those upper-level managers or staff groups developing the survey may have difficulty reaching consensus about the purposes of the survey, its content, or the postsurvey procedures for analyzing, interpreting, and using the results. A related problem may be that supervisors and/or unions may be so committed to existing policies and practices that the survey may be seen by employees as a meaningless exercise.

Another potential disadvantage is organizational disturbance. The survey may either call attention to quiescent issues or induce unrealistic expectations about subsequent actions; either consequence potentially could disrupt normal operations.

Finally, although survey data may help in recognizing relationships in data and suggest ways of improving functions, they do not actually establish cause-effect relationships. There is the potential danger that people will assume that correlation implies causation when, in fact, the relationship may not be causal in nature.

Macroergonomics applications. Organizational questionnaire survey data can be very useful in quickly and inexpensively identifying symptoms of work system design problems and where in the work system those problems may be occurring (e.g., in a specific unit of the organization or throughout the entire work system). Sometimes problems may be identified in one aspect of the work system and a questionnaire developed and used to see how widespread the problem is throughout the organization.

In addition, when used in its survey feedback form as an integral part of the participatory ergonomics process, the survey can provide both managers and employees with data to help them in identifying work system design problems, and/or suggest what needs to be done to correct or improve the work system's functioning. Finally, when used as a pre- and postmacroergonomics intervention measure, the questionnaire can assist in evaluating the effect of work system redesign efforts and suggest where further redesign may be needed.

Interview Surveys

Another frequently used method for gathering information in organizations is the interview, typically the *stratified, semistructured interview survey*. The interview is stratified in that a representative sample of all levels and units within the employee groups of interest are interviewed. Because both time and expense usually make interviewing all employees (or even a large sample of them) impractical, a comparatively small, stratified sample offers a practical means of ensuring that the data gathered will be representative of the full group of interest. A *semistructured* approach means that the interviewer asks certain key questions that systematically tap the topics of interest, but then he or she can ask follow-up questions ad hoc, depending on the response received for each of the key questions.

Advantages. One major advantage is that the interviewer can observe both oral and nonverbal responses to questions. Sometimes the nonverbal responses provide more important data than the oral. Second, unlike questionnaire surveys, the interviewer can follow up on initial responses and explore unanticipated responses or unanticipated issues, or go into a given issue in greater depth. Finally, if the interviewer establishes a good rapport with the interviewee, the interviewee will become more ego involved in the process than if responding to a comparatively impersonal written questionnaire. Respondents can put more thought and effort into their responses.

Disadvantages. The major disadvantages of interviews are time and cost – the time and costs of training or hiring skilled interviewers and costs associated with conducting the interviews. Whereas a written questionnaire survey can be administered to a large group at once, interviews must be done with one or a very few persons at a time. Further, an interview to gather organizational data typically takes an hour or more.

A second major disadvantage is that interviewees lose their anonymity. As a result, they may be inclined to "play it safe" and only give sociably acceptable answers or answers they think the interviewer wants to hear. This disadvantage can be minimized by (a) using skilled, trained interviewers, (b) interviewing employees on their own "turf" rather than in the interviewer's or manager's office, (c) clearly explaining the purpose of the interview and what will be done with the results, and (d) assuring the interviewee that his or her responses will be kept confidential.

Macroergonomics applications. Semistructured interview surveys can be a particularly effective way of identifying and gaining insights into problems with a current work system's design. They can also reveal specific kinds of macroergonomics intervention that might be effective in either redesigning the work system or implementing the redesign. The interview survey approach can be highly useful in identifying incompatibilities between the macro-design of the work system and the micro-design of individual jobs and/or related human-machine and human-software interfaces.

Focus Groups

Essentially, the focus group method is a variation of the interview survey approach and partially overcomes the interview's disadvantages of time and money. At the same time, it retains the basic advantages of the interview method. Fundamentally, the focus group approach involves bringing people together to be interviewed as a group about one or more specific issues.

Advantages. In addition to the savings in time and money as compared with the traditional interview survey method, the focus group method has one other important advantage: Hearing a colleague's responses may stimulate the thinking of other participants, thus yielding additional useful data that otherwise might not have surfaced.

Disadvantages. In addition to the loss of anonymity, characteristic of all interview approaches, a focus group participant's responses will be heard by the other participants. Others may then be reluctant to express ideas or perceptions that will not be accepted or popular with the group. In addition, there always is the possibility that a "groupthink" phenomenon will occur. (Groupthink arises when pressure for consensus among group members inhibits the expression of opinions that threaten consensus and leads them to collectively construct rationalizations that discount warnings or problems and, ultimately, to the illusion of unanimity of opinion.)

Macroergonomics applications. When applied within the context of macroergonomics, the focus group brings people from a particular work system together to be interviewed about specific aspects of the work system, or its sociotechnical environment, which are of interest to the macroergonomist. It may be conducted as a semistructured interview. Sometimes a work system change is simulated, and group members are then interviewed together to gain their collective perceptions or opinions about specific aspects of the change. Various other data methods also may be used (e.g., *Monte Carlo technique*, described in the Glossary, or paired comparisons, in which each item is compared with every other item on some attribute and, each time, one is chosen over the other).

Combining Methods

As is suggested throughout this chapter, often two or more of the techniques described are used together in carrying out a macroergonomics analysis, intervention, and evaluation. For example, a participatory ergonomics approach may be used throughout the entire macroergonomics process, but it also could involve

a. conducting a field study to observe the existing work system and related performance measures,
b. doing a laboratory simulation study of a proposed work system change with some employees,
c. conducting a field experiment in which a part of the work system is modified and performance effects observed,
d. using focus groups to evaluate the field experiment, and
e. conducting an organizational questionnaire survey after implementing the work system redesign to obtain employee responses to the change.

In addition, there are other methods that are somewhat unique to macroergonomics. Two of these, sociotechnical analysis and design of work system *structure* and sociotechnical analysis and design of work system *process*, will be described in Chapters 4 and 5. These can be, and usually are, used in combination with the methods described in this chapter.

A number of other unique methodologies have been developed for use in specific macroergonomics applications. For example, from a review of the proceedings for the first three International Ergonomics Association International Symposia on Human Factors in Organizational Design and Management (Brown & Hendrick, 1986; Hendrick & Brown, 1984; Noro & Brown, 1990), the following technologies were identified (Hendrick, 1991):

- User systems analysis (5 papers)
- Systems analysis modeling (4 papers)
- Ergonomics work analysis (2 papers)
- Work systems design (2 papers)
- Usability test methodology
- Function analysis modeling
- Fuzzy concepts as a macroergonomics tool
- Modified "garbage-can" model for evaluating organizational design alternatives
- Task allocation charting
- Systematic organizational design methodology (SORD) for designing organizational units
- Use of computer-assisted design to simulate an organization
- Organizational requirements definition tools (ORDIT) for assisting in specifying organizational requirements for information technology systems.

These methods have also been used in combination with the methods described in this chapter.

Chapter 4

ANALYSIS AND DESIGN OF WORK SYSTEM STRUCTURE

As noted in Chapter 2, the design of a work system's structure and related processes involves consideration of three major sociotechnical system components that interact and affect optimal work system design: the technological subsystem, personnel subsystem, and relevant external environment that permeates the organization. Each of these components has been studied in relation to its effects on the three organizational design dimensions described in Chapter 1 (complexity, formalization, and centralization) and empirical models have emerged. These models can be used as macroergonomics tools in analyzing and developing or modifying the design of a given work system. Some of the models that have proven particularly useful to the authors are described in this chapter.

Technological Subsystem Analysis

Technology, as a determinant of work system structure, has been operationally classified in several distinctly different ways that are useful to macroergonomics: (a) by the mode of production, or *production technology;* (b) by the action individuals perform on an object to change it, or *knowledge-based technology;* and (c) by the degrees of automation, work-flow rigidity, and quantitative specificity of evaluation of work activities, or *work flow integration.* From each of these empirically derived classification schemes emerges a major generalizable model of the technology-organizational design relationship.

Woodward: Production Technology

Joan Woodward and her associates (1965) were the first to study technology as a determinant of organizational structure. They were looking for differences between successful and less successful organizations within the same industry, across a variety of industries. These researchers studied 100 manufacturing firms in South Essex, England, with at least 100 employees. The companies varied in size, managerial levels (2–12), span of control (2–12 at the top; 20–90 at the first-line supervisory level), and ratio of line employees to staff personnel (less than 1:1 to more than 10:1). Using interviews, systematic observations, and review of company records, Woodward et al. noted the following factors (among others) for each firm: (a) the organization's mission and significant historical events, (b) the manufacturing processes and methods utilized, and (c) the organization's success, including changes in market share, relative growth or stagnation within its industrial field, and fluctuation of its stock prices.

From this analysis, Woodward et al. found that the industries they studied could be classified in terms of their modes of technology: *unit, mass,* or *process* production. These modes were conceptualized as representing categories on a scale of increasing technological complexity. At the least complex end were the unit and small batch producers that produce custom-made products, such as hand-crafted pottery or jewelry. Next were the large batch or mass production firms. These companies produced automobiles and other more or less standardized products and used predictable, repetitive production steps. Highest in production complexity were heavily automated process production firms, such as oil and chemical refineries.

Within each type of production mode, Woodward and her colleagues found that all that discriminated between the successful and less successful firms were the characteristics of their organizational structure. Specifically, three organizational structure variables were found to increase as technological complexity increased. First, as *technological complexity increased, the degree of vertical differentiation also increased*. The successful firms within each technology mode tended to have the median number of hierarchical levels for that category. In Woodward et al.'s sample, this optimum number of levels for unit producers was three, for mass it was four, and for process, six. The less successful firms within each production category had a greater or lesser number of levels.

Second, *as technological complexity increased, the optimal ratio of administrative support staff to industrial line personnel increased*. The findings for the successful firms in each technology category were as follows:

- Unit production firms had low complexity with little line and staff differentiation, first-line supervisors had relatively narrow spans of control, jobs were widely rather than narrowly defined, and formalization and centralization were low.
- Mass production units had high complexity with clear line and staff differentiation, first-line supervisors had relatively broad spans of control, jobs were narrowly defined, and formalization and centralization were high.
- Process production units had high vertical differentiation with little line and staff differentiation, supervisors had wide spans of control, and formalization and centralization were low.

Several follow-up studies have lent support to Woodward et al.'s findings (Harvey, 1968; Zwerman, 1970). However, it is important to note that Woodward et al. imply causation when, in fact, their methodology really only establishes correlation. Also note that their data were collected from within a single culture and at a particular point in time. In a different culture, or at some other time, the psychosocial, cultural, and other environmental factors might conceivably result in somewhat different interactions with production mode in terms of their influence on work system design.

Perrow: Knowledge-Based Technology

Although it can be a useful analytical tool for macroergonomics, a major shortcoming of Woodward et al.'s model is that it applies only to manufacturing firms, which constitute less than half of all organizations. A more generalizable model of the technology-work system structure relationship, developed by Perrow (1967), uses a *knowledge-based* scheme rather than a production classification scheme. Perrow starts by defining technology as the action performed on an object in order to change it. This action requires some form of technical knowledge. Using this approach, he identified two underlying dimensions of knowledge-based technology. The first is *task variability*, or the number of exceptions or nonroutine problems encountered in one's work. The second concerns the type of search procedures available for responding to task exceptions, or *task analyzability*.

These search procedures can range from "well defined" to "ill defined." At the "well-defined" end of the continuum, solving problems can be accomplished using rational-logical, quantitative, and analytical reasoning. At the "ill-defined" end, there are no readily available formal search procedures, and one must rely on experience, judgment, and intuition. Dichotomizing these two dimensions yields a two-by-two matrix with four cells. Each cell represents a different knowledge-based technology (see Table 4.1).

TABLE 4.1
Perrow's Classification Scheme (matrix)

		Task Variability	
		Routine with few exceptions	High variety with many exceptions
Problem Analyzability	Well defined and analyzable	Routine	Engineering
	Ill-defined and unanalyzable	Craft	Nonroutine

Routine technologies have well-defined problems with few exceptions. Mass production units typically fall into this category (e.g., automobile or television assembly plants). Routine technologies lend themselves to standardized coordination and control procedures and, accordingly, are associated with high formalization and centralization.

Nonroutine technologies have many exceptions and problems that are difficult to analyze. Combat aerospace operations, such as air-to-air combat, are an example. These technologies require flexibility; accordingly, they should be highly decentralized and have low formalization.

Engineering technologies (e.g., mechanical, electrical) have many exceptions but can be handled using well-defined rational-logical processes. They thus lend themselves to moderate centralization but require the flexibility that is achievable through low formalization.

Craft technologies (which create, for example, art objects or Web pages) involve fairly routine tasks, but problem solving relies heavily on the experience, judgment, and intuition of the craftsperson. Accordingly, decisions must be made by those with expertise, which requires decentralization and low formalization.

Perrow's model has been supported by considerable empirical research in both the private and public sectors (e.g., Hage & Aiken, 1969; Magnusen, 1970; Van de Van & Delbecq, 1979). We have found this model to be particularly useful for assessing an organization's technology to determine its implications for the firm's work system structure.

Work-Flow Integration

After studying a wide range of manufacturing and service organizations, a team of researchers at the University of Aston (UK) concluded that technology can be defined in terms of three basic characteristics: automation of *equipment,* or the extent to which work activities are performed by machines; *work-flow rigidity,* or the extent to which the sequence of work activities is inflexible; and *specificity of evaluation,* or the degree to which work activities can be assessed by specific, quantitative means. The Aston group found these characteristics to be highly related, so they combined them into a single scale labeled *work-flow integration* (Hickson, Pugh, & Pheysey, 1969). Within smaller organizations (1000 or fewer employees), work-flow integration was found to be weakly related to organizational structure. In general, as work-flow integration increases, specialization, formalization, and decentralization of tactical authority also increase for optimal functioning. This relationship was not as apparent in larger organizations with thousands of employees.

The most important outcome of the Aston studies was that although technology was found to affect organizational structure, it appears to have significantly less impact than the other two sociotechnical system elements: the personnel subsystem and relevant external environment. Of special importance for macroergonomics, the Aston studies further demonstrated that the so-called technological imperative – the view that technology has a compelling influence on work system structure and thus should determine work system design – clearly overstates the case (Baron & Greenberg, 1990). As noted in Chapter 2, this finding was first documented in the Tavistock Institute studies of the longwall method of coal mining. In spite of these conclusive findings, the myth of technological determinism continues to persist.

Other Technological Considerations

Over the past several decades, major advances have been made in computer and communications technology. Two forms of this technology have major implications for work system design: *advanced information technologies* (AIT) and *computer-integrated manufacturing* (CIM). AIT has tended to facilitate the efficiency of decentralizing operational or tactical decision making while enhancing the efficiency of centralized strategic decision making (Bedeian & Zammuto, 1991). In other words, AIT links employees electronically, thus better enabling them to participate in the tactical decision-making process. As a result, AIT enhances the efficiency of decentralization and fosters greater professionalism. AIT also enables lower-level personnel to have a greater *indirect* influence on strategic decision making; this happens because they often are the ones who select and filter the information and structure the databases that form the input for the highly centralized strategic decisions.

CIM, by its very nature, results in a very high level of integration of work-flow processes and, consequently, a high level of interdependence among differentiated units. A major result of this interdependence is an increased need for effective integrating mechanisms across functional units. In addition, CIM often increases the need for market-based unit grouping (e.g., product task team grouping), especially during the product design phase (Bedeian & Zammuto, 1991; Drucker, 1988).

Personnel Subsystem Analysis

At least three major characteristics of the personnel subsystem are critical to an organization's work system design: (a) the *degree of professionalism*, (b) *demographic characteristics*, and (c) *psychosocial aspects* of the workforce (Hendrick, 1997).

Degree of Professionalism

Robbins (1983) noted that formalization can take place either on or off the job. When done on the job, formalization is *external* to the employee. The rules, procedures, and human-system interfaces are designed to limit employee decision discretion. This situation typically characterizes unskilled and semiskilled positions and is what is meant by the term *formalization*. Professionalism, on the other hand, creates *internal* formalization of behavior through a socialization process that is an integral part of education and training. Through formal education and training, employees learn the values, norms, and expected behavior patterns of the job before entering the organization.

Thus, from a macroergonomics standpoint, there is a trade-off between formalization of the work system and professionalization of the jobs in the work system and job design process. Where the work system is designed to allow for low formalization and, thus, considerable employee discretion, jobs need to be designed to require people with a relatively higher level of professional training or education. In the absence of formal decision rules and procedures, employees need to have the necessary professional knowledge to make decisions. Often, it is the need to have employees who can deal with unique or unanticipated situations that creates the need for low formalization and more highly professionalized jobs.

Demographic Factors

Demographic characteristics of the workforce that will (or do) comprise the organization's personnel subsystem can potentially interact with the work system's design. The most striking examples in the United States are (a) the "graying" of the workforce, (b) value system shifts, (c) the broadening of the cultural diversity of the workforce, and (d) the increased number of women in the workforce.

Graying of the workforce. In the United States and elsewhere, as the post-World War II baby boomers move through their working careers, the average age of the workforce has been increasing at the rate of about 6 months per year. This trend began in the late 1970s and continued through the 1990s, resulting in an older, more experienced, more mature, and better-trained workforce. As a result of this shift, the workforce has become more professionalized.

If employees are to feel fully utilized and remain motivated toward their work, organizational designs need to adapt to this change by becoming less formalized and decentralizing more of the decision making. Work system designers need to consider these factors carefully, particularly in designing functions and jobs in which high formalization and centralization traditionally have characterized the structure of the work system.

Value system shifts. After considering findings from extensive longitudinal studies of workforce attitudes and values, Yankelovich (1979) noted that workers born after World War II have very different views and feelings about work than did their predecessors. Furthermore, he said, these conceptions and values have a profound effect on work systems. This group of workers, referred to as the "new breed" by Yankelovich, have three principal values that differ from values held by older workers: the increasing importance of leisure, the symbolic significance of the paid job, and the insistence that jobs become less depersonalized and more meaningful.

According to Yankelovich, when asked what aspects of work are more important, the new-breed employee stresses "being recognized as an individual" and "the opportunity to be with pleasant people with whom I like to work." From a work system design standpoint, these values translate into a need for less hierarchical, less formalized, and more decentralized organizational structures. Attendant on these needs is the need to design greater professionalism into individual jobs and human-system interfaces than are found in traditional bureaucracies. Operationally, these design characteristics allow for greater individual recognition and respect for an employee's worth. In other national opinion survey research, Yankelovich (1988) found majority views on a variety of issues that are consistent with new-breed values.

One result of movement of the World War II workforce out of organizations and the inward movement of the new breed has been increasing demand by workers for better-designed jobs, greater participation in decision making, and less formalization.

Cultural diversity. In a number of urban U.S. areas, such as the Los Angeles basin, immigration has resulted in the development of workforces that are far more culturally diverse than existed several decades ago. It has become apparent that unless organizations adapt to this diversity, they will experience adverse effects on employee motivation and commitment (Jackson, 1992; Thomas, 1991). Much of the adaptation required involves changing organizational cultures to be more inclusive. Decentralizing some aspects of decision making to allow employees greater control in their work groups appears to be one structural change that can facilitate this process. This action should involve participatory ergonomics in designing or modifying work systems.

Women. During the 1980s and 1990s, women entered the workforce in greater and greater numbers. As yet, there is no clear indication of how these demographic changes will – or should – affect work system design. In our consulting work we have seen that as women have moved into positions traditionally held by men, they have tended to emphasize the importance of modifying work systems and jobs to allow for greater social interaction, particularly when opportunity for social interaction has been limited.

Psychosocial Factors

Harvey, Hunt, and Schroder (1961) identified the higher-order structural personality dimension of concreteness-abstractness of thinking, or *cognitive complexity,* underlying different conceptual systems for understanding reality. In general, the more a given culture or subculture encourages, by its child-rearing and educational practices, exposure to new experiences or diversity, and the more it provides opportunities for exposure to diversity through affluence, education, communications media, transportation systems, and so forth, the more cognitively complex people will become. Active exposure to diversity increases a person's opportunity to develop new conceptual categories and sub-categories in which to store experiences (referred to as *differentiation*). People who are actively exposed to diversity also learn new rules and combinations of rules for *integrating* information and deriving deeper, more insightful conceptions of problems and solutions. As a result, those actively exposed to considerable diversity develop a high degree of conceptual differentiation and integration, or *abstract functioning*.

Conversely, relatively closed-minded approaches to new experiences or a lack of exposure to considerable diversity results in a more limited development of differentiation and integration in a person's conceptualizing of reality, or *concrete functioning*.

Concrete functioning has been found to be characterized by a relatively high need for structure, order, stability, and consistency; a low tolerance for ambiguity; a closed belief system; absolutism; authoritarianism; paternalism; and ethnocentrism. Concrete persons tend to interpret their world more literally and statically than do abstract persons. They tend to see their views, values, norms, and institutional structures as relatively static and unchanging.

Conversely, abstract adult functioning is characterized by a low need for structure, order, stability, and consistency; a high tolerance for ambiguity; openness of beliefs; relativistic thinking; a greater capacity for empathy; and strong "people orientation." Abstract people have a dynamic conception of their world and *expect* their views, values, norms, and institutional structures to change (Harvey, 1963; Harvey et al., 1961; Hendrick, 1979, 1981, 1996).

Hendrick (1979, 1981, 1990) found evidence suggesting that relatively concrete work groups and managers function best in mechanistic organizations characterized by moderately high vertical differentiation, centralization, and formalization and in which the work system structure and processes are unambiguous and relatively slow to change. In contrast, although they can perform well in mechanistic organizations, cognitively complex or abstract people prefer more organic organizational designs characterized by relatively low levels of vertical differentiation, formalization, and centralization.

Personnel Subsystem Implications for Work System Design

Much of the available data on personnel subsystem determinants of work system design are in the form of either attitude survey results or projections from psychosocial and demographic studies. Despite their somewhat tenuous nature, these data converge: They indicate a need for organizations of the future to be as vertically undifferentiated, decentralized, and lacking in formalization as their technology and environments will permit. Given the rapidly developing trend toward highly dynamic virtual organizations and attendant work systems, this indication is most fortunate.

External Environmental Characteristics

The very survival of organizations depends on their ability to adapt to their external environment. In terms of open systems theory, organizations require monitoring and feedback mechanisms to follow and sense changes in their relevant task environments and a capacity to make responsive adjustments. *Relevant task environments* refers to that part of the firm's external environment that can positively or negatively influence the organization's effectiveness – the organization's critical constituencies.

Types of External Environments
Examining field studies of 92 industrial firms in five countries, Negandhi (1977) identified five types of external environments that significantly affect organizational functioning:

1. *Socioeconomic* – particularly the degree of stability of the socioeconomic environment, nature of the competition, and availability of materials and qualified workers.
2. *Educational* – the availability of facilities and programs for employees or potential employees in the local region and the educational level and aspirations of workers.
3. *Political* – the degree of stability at all governmental levels and the government's attitudes toward business (friendliness versus hostility), labor (friendliness versus hostility) and control of prices.
4. *Cultural* – social status and caste system in the community; values and attitudes of employees and their families toward work, management, and so on; and the nature of trade unions and union-management relationships.
5. *Legal* – degree of legal controls, restrictions, and compliance requirements.

The relevant task environments are different for each organization in type, qualitative nature, and importance. The particular weighted combination of relevant task environments for a given organization constitutes its *specific task environment*. A major determinant of an organization's specific task environment is its *domain*, or the range of products or services offered and market share (Robbins, 1983). Domain is important because it determines the points at which the organization depends on its specific task environment (Thompson, 1967).

A second determinant of an organization's specific task environment is its *stakeholders*. These include the firm's stockholders, lenders, members of the organization, customers, users, governmental agencies, and the local community(s). Each has an interest in the organization.

Environmental Uncertainty

Of particular importance to work system design is the fact that all specific task environments vary along two highly critical dimensions: *change* and *complexity* (Duncan, 1972). Degree of change refers to the extent to which a given task environment is dynamic as opposed to remaining stable over time. The degree of complexity refers to whether the components of an organization's specific task environment are many as opposed to few in number (i.e., does the company interact with few or many government agencies, customers, suppliers, and competitors?). These two environmental dimensions of change and complexity combine to determine the *environmental uncertainty* of an organization.

Table 4.2 illustrates this relationship for four levels of uncertainty and provides the environmental characteristics and a representative industry for each.

TABLE 4.2
Environmental Uncertainty Dimensions (matrix)

		Degree of Change	
		Stable	Dynamic
Degree of Complexity	Simple	Low uncertainty (e.g., education)	Moderately high uncertainty (e.g., retail)
	Complex	Moderately low uncertainty (e.g., military)	High uncertainty (e.g., software)

Two major generalizable models have been empirically derived for assessing environmental uncertainty as a determinant of organizational structure. One model focuses directly on environmental uncertainty; the other treats environmental uncertainty as one (albeit the most important) of several key environmental dimensions affecting work system structure.

Burns and Stalker: Environmental uncertainty. Based on studies of 20 English and Scottish industrial firms, Burns and Stalker (1961) found that the type of work system structure that worked best in a relatively stable and simple organizational environment was very different from that required for a more dynamic and complex environment. For stable, simple environments, *mechanistic* structures worked best. Mechanistic work systems are characterized by high vertical and horizontal differentiation, formalization, and centralization. They typically have routine tasks and programmed behaviors and cannot respond to change quickly. A strong emphasis is placed on stability and control.

Conversely, for dynamic, complex environments, *organic* structures worked best; these are characterized by flexibility and quick adaptability. Organic work systems emphasize (a) lateral rather than vertical communication, (b) influence based on knowledge and expertise rather than position and authority, (c) information exchange rather than directives from above, (d) conflict resolution by interaction rather than by superiors, and (e) relatively loosely defined responsibilities. Accordingly, organic work systems have low vertical differentiation and formalization and decentralized tactical decision making. Similar findings were implicit in Emery and Trist's (1965) analysis of the effects of environmental instability on sociotechnical systems.

Lawrence and Lorsch: Subunit environment and design complexity. A common characteristic of complex specific task environments is that organizations usually develop specialized units to deal with particular parts of the environment. Lawrence and Lorsch (1969) conducted field studies to determine what type of work system design was best for coping with different economic and market environments. They studied companies in various industries (e.g., food, plastics, and containers), which varied considerably in their degree of environmental uncertainty. Based on their studies, Lawrence and Lorsch identified five major variables that can be assessed regarding subunit environments to determine the optimal degree of horizontal differentiation: (a) *uncertainty of information* (low, moderate, high), (b) *time span of feedback* (short, medium, or long), (c) pattern of *goal orientation* (focus of tasks), (d) pattern of *time orientation* (short, medium or long), and (e) pattern of *interpersonal relationships* (task or social). In general, the more dissimilar the functions on one or more of these dimensions, the stronger the likelihood that the functions should be differentiated into separate subunits (departmentalized) for effective functioning.

Lawrence and Lorsch also found that the greater the differentiation, the greater the need for integrating mechanisms, and that the level of environmental uncertainty was of foremost importance in selecting the structure appropriate for effective functioning. Subunits with more stable environments (e.g., production) tended to have high formalization, whereas those operating in less predictable environments (e.g., research and development) had low formalization.

Lawrence and Lorsch's research is particularly important to macroergonomics because it demonstrates that whenever an organization's design does not fit its mission, external environment, or resources, its functioning is likely to suffer.

Integrating the Results

The separate analyses of the key characteristics of the organization's technological subsystem, personnel subsystem, and specific task environment provide results about the structural design of the work system. In many cases, these results show a natural convergence, but sometimes the outcome of one of the analyses may be at variance with the others. When this occurs, the macroergonomics professional must reconcile the differences.

Based on suggestions from the literature and one of the author's (Hendrick) personal experiences, the outcomes from the analyses can be integrated by weighting them approximately as follows: If the technological subsystem analysis is given a weight of 1, the personnel subsystem analysis should be assigned a weight of 2 and the analysis of the specific task environment a weight of 3. For example, let's assume the technological subsystem characteristics indicate a more formalized and somewhat centralized work system design (a weight of 1), the personnel subsystem analysis indicates an intermediate level of professionalism and associated moderately low level of formalization (a weight of 2), and the specific task environment analysis indicates a need for a low level of formalization and centralization and related high level of professionalism (a weight of 3). Combining these weighted outcomes would indicate that one should choose a work system design characterized by a fairly high level of professionalism and fairly low level of formalization and centralization.

In applying the results of the analyses, it is important to note – as pointed out in the aforementioned research of Lawrence and Lorsch – that specific functional units of an organization may differ in the characteristics of their technology, personnel, and specific task environments, particularly within larger organizations. Accordingly, the separate functional units may themselves need to be analyzed as though they were separate organizations. As a result, the design of the work system suggested by the results may differ for the different functional units.

Consideration of Job Design Characteristics in Macroergonomics

In macroergonomically designing or modifying a work system, it is important to be continually aware of the impact that decisions about work system structure (and processes) are likely to have on the design of individual jobs. Hackman and Oldham (1975) empirically identified five job characteristics that are critical to intrinsic job motivation, employee self-worth, stress reduction, and satisfaction for growth-oriented employees: (a) task *variety* (having different meaningful things to do in one's work), (b) *identity* (sense of job wholeness), (c) *significance* (perceived job meaningfulness), (d) *autonomy* (control over one's work), and (e) *feedback* (knowledge of results). Without these characteristics, jobs are often dehumanizing, are less psychologically meaningful, reduce employees' sense of responsibility; and frequently appear to lead to high stress, demotivated employees, job dissatisfaction, absenteeism, and reduced productivity (Organ & Bateman, 1991).

Since the time these job characteristics were first identified, their importance has been noted repeatedly in many types of organizations and work situations. For example, Bammer (1990, 1993) conducted a meta-analysis of the field studies of upper-extremity work-related musculoskeletal disorders among computer operators reported from around the world during the 1980s. She found no consistent relationship of nonwork factors to employee musculoskeletal disorders. The data on biomechanical factors led her to conclude that these factors are important, and (micro-ergonomic) efforts to improve them should be encouraged.

Bammer also noted that, by themselves, biomechanical improvements are an insufficient means to reduce work-related musculoskeletal disorders. The factors that did consistently relate to musculoskeletal disorders across studies were work organization variables. Bammer concluded that "improvements in work organization to reduce pressure, and to increase task variety, control, and the ability for employees to work together must be the main focus of prevention and intervention"; furthermore, "ironically, such improvements in work organisation generally also lead to increased productivity" (1993, p. 35). Put simply, what Bammer identified as the key correlates of work-related musculoskeletal disorders were Hackman and Oldham's job characteristics plus the opportunity to satisfy social needs on the job.

Based on the literature and our personal consulting experiences, we recommend that when making design decisions about the work system, do not preclude designing individual jobs with Hackman and Oldham's five job characteristics. We also recommend that attention be paid to ensuring that the work system design enable employees to satisfy social needs on the job.

Selecting the Right Structural Form

From a macroergonomics perspective, structural analysis involves more than considering how sociotechnical system variables should shape the basic dimensions of the work system; it also involves integrating these dimensions into an overall *structural form*. To this end, a variety of types of structural form are available to system designers.

As is the case when designing the individual dimensions of work system structure, the different types of structural form can enhance or inhibit organizational functioning, depending on the organization's specific sociotechnical characteristics. The key, then, is to select the structural form for the work system that best fits the organization's sociotechnical characteristics and related work system dimensions. In this section we consider the four general types of organizational structure most commonly found and discuss the advantages and disadvantages of each. We include guidelines for determining when each type is, and is not, likely to be appropriate.

The four general types of overall organizational structure are (a) classical *machine bureaucracy*, (b) *professional bureaucracy*, (c) *matrix organization*, and (d) *free-form design* (Robbins, 1983). Note that large, complex organizations often have relatively autonomous units with different overall forms. In general, the larger the organization, the greater the likelihood that it will utilize more than one type of work system structure.

Classical or Machine Bureaucracy

This form of work system has its roots in two streams of thought: *scientific management* and the *ideal bureaucracy*.

Scientific management. By the end of the nineteenth century, industrial technology was rapidly developing in American and European industry. Labor was becoming highly specialized, and engineers were called on to help design organizations and optimize efficiency. One of these engineers, Frederick W. Taylor (1911), through his concept of scientific management, has had a major impact on the shaping of classical organizational theory. Taylor's concepts of work system organization are implicit in his four basic principles of managing:

First. Develop a science for each element of man's work that replaces the old rule-of-thumb method.

Second. Scientifically select and train, teach, and develop the workman. In the past he chose his own work and trained himself as best he could.

Third. Hardily cooperate with the men in order to ensure all of the work is being done in accordance with the principles of the science that has been developed.

Fourth. Provide equal division of work and responsibility between the management and the workmen. The management takes over all work for which they are more qualified than the workmen. In the past, almost all the work and the greater part of the responsibility were thrown upon the men. (Szilagyi & Wallace, 1990, p. 662)

Ideal bureaucracy. The classical bureaucratic design was conceptualized by Max Weber. Weber recommended that organizations adhere to the following work system design principles:

1. All tasks necessary to accomplish organizational goals must be divided into highly specialized jobs. A worker must master his

trade, and this expertise can be more readily achieved by concentrating on a limited number of tasks.

2. Each task must be performed according to a "consistent system of abstract rules." This practice allows the manager to eliminate uncertainty due to individual differences in task performance.

3. Offices or roles must be organized into hierarchical structure in which the scope of authority of superordinates over subordinates is defined. This system offers the subordinates the possibility of appealing a decision to a higher level of authority.

4. Superiors must assume an impersonal attitude in dealing with each other and subordinates. This psychological and social distance enables the superior to make decisions without being influenced by prejudices and preferences.

5. Employment in a bureaucracy must be based on qualifications, and promotion is to be decided on the basis of merit. Because of this careful and firm system of employment and promotion, it is assumed that employment will involve a lifelong career and loyalty from employees. (Weber, 1946. p. 214)

Weber believed that strict adherence to these organizing principles was the "one best way" to achieve organizational goals. He assumed that if organizations implemented a structure that emphasized administrative efficiency, stability, and control, they could obtain optimal efficiency (Szilagyi & Wallace, 1990).

Collectively, these theoretical principles of Taylor and Weber resulted in what is known as the *machine bureaucracy* type of organizational structure. Its basic structural characteristics are as follows (Robbins, 1983):

1. *Division of labor.* Jobs are narrowly defined and consist of relatively routine and well-defined tasks.

2. *A well-defined hierarchy.* A fairly tall, clearly defined, formal hierarchical structure in which each lower office is under supervision and control of a higher one. Tasks tend to be grouped by function. Line and staff functions are clearly identified and kept separate.

3. *High formalization.* Extensive use of formal rules and procedures to ensure uniformity and regulate employee behavior.

4. *High centralization.* Decision making is reserved for managers, with relatively little employee decision discretion.

5. *Career tracks for employees.* Members are expected to pursue their careers within the organization; career tracks form part of the work system design for all but the most unskilled positions.

Advantages. The primary advantages of the machine bureaucracy are administrative efficiency, stability, and control over the organization's functioning. Having narrowly defined jobs with a clear set of routinized tasks minimizes the likelihood of error, better enables employees to know their own function and those of others, requires comparatively few prerequisite skills, and minimizes training time and costs. Formalization ensures stability, control, and a smooth, integrated pattern of functioning. Centralization further ensures control and enhances stability.

Disadvantages. The machine bureaucracy form of work system has at least two major disadvantages. First, machine bureaucracies are inherently slow and inefficient in responding to environmental change and nonroutine situations. Second, they tend to result in jobs that fail to make adequate use of workers' mental and psychological capacities. Consequently, jobs tend to lack intrinsic motivation.

When to use. The machine bureaucracy form of work system should be used when the relevant external environments are comparatively simple, stable, and/or predictable; employees' education and skill levels are relatively low; and system operations tend to be repetitive or otherwise can be routinized. Where these conditions do not exist, one of the other forms of overall organizational structure is likely to be more effective.

Professional Bureaucracy

The professional bureaucracy design relies on a relatively high degree of professionalism in the jobs that constitute the work system. It differs from the machine bureaucracy in that jobs are more broadly defined, less routine, and allow for greater employee decision discretion (Robbins, 1983). Consequently, there is less need for formalization, and tactical decision making is decentralized. In addition, fewer levels of hierarchy may be needed. As in machine bureaucracies, positions are grouped functionally and are hierarchical, and strategic decision making often remains centralized.

Advantages. Compared with machine bureaucracies, the professional bureaucracy has at least three major advantages. First, it can cope more efficiently with complex environments and nonroutine tasks. Second, jobs tend to better utilize employees' mental and psychological capabilities, so they are more intrinsically motivating. Third, it requires less managerial tactical decision making and control, thus freeing managers to give greater attention to long-range planning and strategic decision making.

Disadvantages. Professional bureaucracies are not as efficient as machine bureaucracies for coping with simple, stable environments. They require a more highly skilled workforce and attendant increase in training time and expense. Control is less tight, and the distinction between line and staff functions is likely to be less clear. The management skills required by professional bureaucracies also tend to be more sophisticated (i.e., a greater reliance on tolerance for ambiguity and the use of persuasion and facilitation skills, rather than on a simple and direct authoritarian style).

When to use. A professional bureaucracy form of work system exists when the external environment is fairly complex, somewhat unstable or unpredictable, and there is an available pool of professionalized workers. A professional bureaucratic form is less optimal for highly repetitive operations with a simple, stable environment, or if the available management pool is highly authoritarian and concrete in its functioning.

Adhocracies

A major disadvantage of bureaucracies is that they tend to be inefficient in responding to highly complex or dynamic relevant external environments. This is particularly true of machine bureaucracies. Because of this shortcoming, two more recent forms of organization have evolved: *matrix* and *free-form* designs, also known as *adhocracy* designs. An adhocracy can be defined as a "rapidly changing adaptive, temporary system organized around problems to be solved by groups of relative strangers with diverse professional skills" (Bennis, 1969, p. 45).

Structurally, adhocracies are characterized by moderate complexity, low formalization, and decentralization (Robbins, 1983). Adhocracies are staffed by highly skilled employees. Accordingly, horizontal differentiation tends to be high and vertical differentiation moderate to low. This relatively low vertical differentiation reflects low formalization, decentralized decision making, and the low need for supervision because of the high level of professional staffing. Instead of the tight administrative control afforded by centralized decision making and formal rules and procedures, flexibility and rapidity of response are emphasized. It is for this reason that adhocracies require a high level of professionalism and elimination of administrative layers.

When the ability to be adaptive and innovative and to respond rapidly to changing situations and objectives is essential, and when these responses require collaboration of people with different specialties, the adhocracy forms of organization are considerably more effective than the bureaucratic forms (Robbins, 1983).

Compared with bureaucracies, adhocracies have at least three major disadvantages. First, because there are no clear boss-subordinate relationships, the lines of authority and responsibility are ambiguous. Thus, conflict is an integral part of adhocracy. The second disadvantage is psychological and sociological stress. Because the structure of teams or units is temporary, work role interfaces are not stable. By comparison, establishing and dismantling human relationships is a slower psychosocial process, which is stressed any time there is significant work system change. Concrete-functioning employees are especially likely to be strained by these stresses and have difficulty coping.

Third, adhocracies lack both the precision and expediency that comes with routinizing functions and structural stability. Only when these losses in administrative efficiency are more than offset by the gains in efficiency of responsiveness and/or innovation should some form of adhocracy design be considered. The two most common forms of adhocracies are the *matrix* and *free-form* designs.

Matrix organizations. The matrix design combines departmentalization by function with departmentalization by project or product line. Like bureaucracies, matrix organizations have functional departments that tend to be lasting. Unlike bureaucracies, members of the functional departments are farmed out to project or product teams as new projects or product lines develop and the combined technical expertise of the various functional departments is required. When the need for a given functional department's professional input to the project is no longer required, or the level of effort decreases, individuals either return to their "home" department or transfer to another team. The project or product manager supervises the team's interdisciplinary effort, but each team member also has a functional department superviser. Thus, the matrix design violates a fundamental design concept of bureaucracy, *unity of command.*

Advantages of matrix organizations. The primary advantage of matrix designs is that they afford the best of two worlds: the stability and professional support of depth afforded by functional departmentalization, and the rapid-response capability of interdisciplinary teams, such as characterize free-form designs.

Disadvantages of matrix organizations. In addition to the disadvantages that characterize all adhocracy designs, the major disadvantage of matrix organizations is that employees must serve two bosses: their functional department head, who tends to focus on the long term and is remote from the team member's immediate tasks; and the project team director, who tends to focus on the short term and matters immediate to the employees' present tasks. Serving two masters with different goals, responsibilities, and time orientations frequently creates conflict and can disrupt organizational functioning.

A second major problem for employees is that when they are kept on a project too long, they may have difficulty keeping technically current and/or may lose contact with their respective functional departments. Both of these consequences can adversely affect their careers.

When to use matrix organizations. The matrix organization is especially well suited for responding to dynamic and complex relevant external environments – conditions in which both rapid interdisciplinary responsiveness and providing for functional depth in individual disciplines are considered essential.

Free-form design. The free form is the newest of our four general types of work system designs. In its pure form, the free-form organization resembles an amoeba, in that its shape continually changes in order to survive (Szilagyi & Wallace, 1990). The central focus of free-form designs is responsiveness to change in highly dynamic, complex, competitive environments.

In free-form designs, a *profit center* arrangement replaces functional departmentalization. Profit centers are managed by teams and tend to be very results oriented. Thus, free-form organizations have very low formalization and low hierarchical differentiation, and decision making is highly decentralized. Heavy reliance is placed on professionalism, which is expressed operationally through participation, internalized formalization, and autonomy.

Free-form work systems resemble matrix organizations in that project teams are created, changed, and disbanded as required to meet organizational goals and problems. Unlike matrix organizations, there is no underlying functional departmentalization or "home" structure; rather, employees "float" from one project team or cost center to another as they are needed. To function effectively, managers and employees alike need a great deal of personal flexibility. A low need for structure and stability, a high tolerance for ambiguity, and the ability to handle change appear essential.

Advantages of free-form designs. Free-form designs have one major advantage over other work system forms: They can respond to highly competitive, complex, and dynamic relevant external environments with great speed and innovation.

Disadvantages of free-form designs. The one major advantage of free-form designs comes at a cost. The free-form organization has all of the disadvantages of matrix adhocracies, only to a greater extent. Conflict, sociopsychological stress, and inherent administrative inefficiency are integral parts of a continuously changing and amorphous organizational structure. It thus requires a very highly professionalized work force to succeed.

When to use free-form designs. A free-form design should be considered whenever the organization's success or survival depends on speed of response and innovation and a highly professionalized workforce is available. These features tend to characterize small to medium-sized high-technology organizations operating in highly dynamic and complex competitive environments, and semiautonomous "outlaw" subunits of large bureaucratic or matrix organizations (e.g., Lockheed's Skunkworks, which designed many of the military aircraft of the last half of the 20th century and did so faster, cheaper, and with many fewer workers than found in conventional aircraft engineering groups).

New Adhocracy Forms
In response to the opportunities afforded by recent technological advances, new variations of the matrix and free-form designs have begun to emerge.

Particularly worth noting are the *modular* and *virtual* types (Dess, Rasheed, McLaughlin, & Priem, 1995).

Modular form. The modular type of organizational structure outsources nonvital functions while retaining full strategic control. Outsiders are used to manufacture parts, handle logistics, and/or perform housekeeping services, maintenance, or accounting activities. Thus, the "organization" is a central hub surrounded by networks of outside suppliers and specialists. What constitutes the organization is dynamic, in that modular parts can easily be added or deleted as the immediate needs require.

Virtual organizations. In contrast to the central hub of the modular organization, the virtual type consists of a continually evolving network of independent companies. Suppliers, customers, and even competitors link to share skills, costs, and access to one another's markets in order to pursue common strategic objectives (Dess et al., 1995). Participating firms may be involved in multiple alliances. In its purest form, a virtual organization need not have a central office or hierarchy. Participating firms may form a virtual organization to attain specific strategic objectives and then disband when those objectives are met.

The major advantage of the virtual organization is that each firm brings a particular set of competencies to the alliance, thus creating a more competitive, yet highly flexible (virtual) entity. In essence, the virtual organization uses a collectivist strategy to cope with environmental uncertainty and to be competitive.

Chapter 5

ANALYSIS AND DESIGN OF WORK SYSTEM PROCESS

In Chapter 4, we said that sociotechnical analysis of work system *structure* is an excellent prerequisite for the sociotechnical analysis and design of work system *processes*. Work system structure is concerned with organizational design constructs; work system process is concerned with the method by which variances are analyzed and design changes are made over time. Although many valid and potentially useful work process improvement methods exist and are in use, we describe a general framework that has strong theoretical support and that integrates ergonomics interface design, function allocation, and other macroergonomics tools. Specifically, the ideas of Emery and Trist (1978) and Clegg, Ravden, Corbertt, and Johnson (1989) are used to prescribe sociotechnical analysis and function allocation associated with function and task design.

Function allocation strategies and methodologies have been under development since the original Fitts (1951) list was proposed to define human versus machine capability and limitation. More recent function allocation methodologies assume the use of a group or team of analysts but do not specify work system design, decision making, or other requirements of such design teams. Clegg et al. (1989) provided requirements for function allocations. Included are the criteria that the allocation methods be (a) systematic (i.e., methodical), (b) criterion-based (i.e., performance oriented), (c) multidimensional (consider both social and technical elements), (d) capable of handling both large and small functions, (e) iterative (focus on continuous improvement), (f) linked to earlier system design decisions, (g) face-valid in an organizational context (i.e., relevant), and (h) promote participation (by operators).

Using these criteria as guidance, the methodology outlined in this book is adapted and modified from the Emery and Trist (1978) sociotechnical analytical model. It can be used to assess both production and nonproduction work processes. Although the methodology can be applied by an inside or outside consultant, it is best used in a participatory framework in which a facilitator or consultant assists organizational personnel in proceeding through the analysis and design steps. Rarely will the macroergonomist have the opportunity to implement a complete top-down process from start to finish. More

Table 5.1
Ten Phases of Work System Process Analysis and Design

Phase	Subsystem(s)
1. Scanning analysis	Environmental/organizational design
2. System type and performance analysis	Technological
3. Technical work process analysis	Technological
4. Variance data collection	Technological
5. Variance matrix analysis	Technological
6. Variance control & role analysis	Personnel
7. Organizational, joint, & function design	Personnel, technological, and organizational design
8. Responsibility perception analysis	Personnel
9. Support system and interface design	All
10. Implement, iterate, and improve	All

likely, a certain step will be performed (for example, micro-ergonomics), the success of which will lead to the opportunity to return to earlier steps to perform a broader analysis. Table 5.1 lists the 10 phases to this general framework. A more detailed description of each phase and its associated steps follows.

A nontraditional work system – the organization and management of a digital library – will be used to illustrate both the 10 phases of the *macroergonomic analysis and design (MEAD)* framework and the kinds of work systems emerging. A digital library is the on-line equivalent of a physical library. Users can "visit" the library on line and browse for and access materials located in various collections.

PHASE 1: Initial Scanning

1.1 Perform mission, vision, principles analysis

1.2 Perform system scan

1.3 Perform environmental scan

1.4 Specify initial organizational design dimensions

The first phase of sociotechnical analysis of work system process is initial scanning, the process of getting a general understanding of the system and its environment. If an evaluation of work system structure has already been performed as described in Chapter 4, then this step can be expedited by referring to the previous analysis.

Here we have combined a number of tools from several domains to provide a general assessment of the external environment. In terms of sociotechnical systems theory, the open organizational system attempts to seek a steady state by adapting to forces or changes that pass through its borders. As noted in Chapter 2 , the external environment may be the most influential in determining the success of the sociotechnical system, so achieving a valid organization-environment fit is important and, in fact, may be the strongest contributor to positive changes in the work system.

Step 1.1: Perform Mission, Vision, Principles Analysis

Mission, vision, and principles provide identity to the organization. These are often called *identity statements* for a company because they characterize the so-called personality of the organization. Often, there is a gap between what the organization professes as its defining characteristics and its identity as inferred from actual behavior. It is instructive to assess the nature and extent of this gap.

First, the formal company identity statements should be identified and evaluated with respect to their components. A *mission* statement describes the organization's purpose or goals and describes its current activity in terms of its products and/or services. "We offer reliable and efficient information services and scholarly collections to users and authors" might be the mission statement for a digital library. The vision is forward looking – what the organization seeks to become in 5 -10 years. "We aim to become the most productive (in terms of works created) and most accessed library in the world" would be the digital library's vision statement.

The *principles* or *values* of the organization are the cultural elements that define the core attributes of the organization. "We value our users and access to information for all groups, and promote collaborative relationships among users, authors, and other stakeholders" emphasizes guiding principles and implies a set of respective behaviors as well. Ultimately, these underlying attributes should drive behavior that can be observed and measured.

In observing behavior, a gap may be perceived between the organization's inferred values (i.e., observed from behaviors and attitudes) and professed values. This gap can be quantified by administering a survey, but rich qualitative information can also be gleaned from interviews. Has the vision been articulated in such a way that employees are genuinely excited about the organization's future? Have jobs been designed so employees at all levels and all points in the processes truly identify with the organization's products and services, or do employees see the organization's mission simply as "to make money"? As Barnard (1938) pointed out, it is destructive to view the organization's purpose as making money.

Finally, if there is a gap between how people behave and how the company expects people to behave, what are the underlying causes? Are managers observed to exhibit the behaviors they expect of others in the organization? Is the reward system misaligned with the organization's objectives? Are new or redesigned interfaces needed? These are some of the questions that can be used in interviews or converted to items in a Likert-type survey instrument for analysis. Such instruments typically have a 5- or 7-point scale ranging from "strongly agree" to "strongly disagree."

Step 1.2: Perform System Scan

Scanning is the process of defining the workplace in systems terms, including defining relevant boundaries. Several tools are available to assist with scanning. We have used the Deming Flow Diagram (Walton, 1986) extensively for this purpose; it works well with both simple (e.g., food store management) and complex (e.g., virtual library) work systems. Deming claimed this approach helped to transform war-torn Japan in the early 1950s into an industrial powerhouse. Here, the organization's mission is detailed in systems terms. As illustrated in the Deming diagram in Figure 5.1 for a digital library work system, system *outputs* are the products or services provided by the organization.

System *inputs* are the resources transformed to produce products and services. *Suppliers* provide the inputs, and *consumers* receive, pay for, and use the outputs. Other "customers" or stakeholders may be beneficiaries of (i.e., have a vested interest in) the system but either do not pay for or directly use products and services. Deming, who focused on statistical quality control, emphasized *internal control*, or those mechanisms or processes that keep transformation processes from spiraling out of control.

Consumer feedback was seen as instrumental to the design and redesign of transformation processes. The system scan also establishes initial *boundaries* or domains of responsibility that exist between the suppliers and customers. As described by Emery and Trist (1978), there are throughput, territorial, social, and time boundaries to consider. *Throughput boundaries* represent the

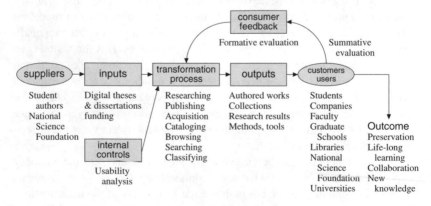

Figure 5.1 Systems Flow Diagram for an on-line digital library work system.

input owned by the system and transformed to output for distribution to consumers. In this case, the internal physical environment is important because it affects work. *Territorial boundaries* have to do with the physical space used for product conversion. Increasingly, this space is distributed and therefore virtual or geographically dispersed, as in the case of a digital library.

Social boundaries are the boundaries created by the formal organization chart's definition of jobs. These are not depicted in a Deming diagram, nor are time boundaries. *Time boundaries* will relate to an organization's mission: seasonality, number of shifts, and the like. In sociotechnical terms, it is instructive to identify variances or deviations from normal processes passed into the production system from suppliers, consumers, and other customers. A list of design changes that could control these variances closer to the source (i.e., inside the organizational system) should be created.

Step 1.3: Perform Environmental Scan

The system scan helps to define the work system's boundaries. Everything outside these boundaries becomes part of the external environment. In the environmental scan, the organization's environmental segments, or *subenvironments*, and the principal stakeholders within these subenvironments are identified. Subenvironments typically include such categories as economic, technological, educational, cultural, and political. Stakeholders within these subenvironments are specific organizations within each category. Their expectations for the organization are identified and evaluated.

Conflicts and ambiguities are seen as opportunities for process or interface improvement. For example, in a digital library system, users might desire access to all available data on a given subject. Librarians, on the other hand, understand that much "data" on the Internet is not valid and therefore is not useful information. The work system then might be interested in some sort of filtering device.

A *realistic future scenario* is created which predicts what would occur if the organization did nothing to change the conflicts and ambiguities. Then, an *idealistic future scenario* is created which defines a future state in which stakeholders' expectations are fully realized. Variances or mismatches between what stakeholders and the work system expect of each other are evaluated to determine design constraints and opportunities for change. The work system itself can be redesigned to align itself with expectations derived from external stakeholders, or, conversely, the work system can attempt to change the expectations of the environment to be consistent with its internal plans and desires.

In our experience, the gaps between work system design and environmental expectations are often gaps of perception. To deal with these gaps, communication interfaces need to be developed between subenvironment personnel and the organization. Specifically, the macroergonomist designs or redesigns interfaces among the organizational system and relevant external subenvironments to improve communication and decision making. These interfaces are referred to as organization- or *work system-environment interfaces.*

Step 1.4: Specify Initial Organizational Design Dimensions

As discussed in Chapter 4, environmental factors are determinants of organizational design, as are factors from the technological and personnel subsystems. Rather than wait to address organizational design until all subsystems have been analyzed, it is useful to develop organizational design hypotheses at this stage based on the environmental scan. By referring to the empirical models of the external environment (e.g., Burns & Stalker, 1961; Lawrence & Lorsch, 1969), we can hypothesize optimal levels of complexity (both differentiation and integration), centralization, and formalization (see Chapter 4).

PHASE 2: Production System Type and Performance Expectations
2.1 Define production system type
2.2 Define performance expectations
2.3 Specify organizational design dimensions
2.4 Define system function allocation requirements

Step 2.1: Define Production System Type

As detailed in Chapter 4, it is important to identify the work system's *production type*, given that the type of production system can help determine optimal levels of complexity, centralization, and formalization. The system scan performed in the previous phase should help in this regard. The analyst should consult the production models discussed previously (i.e., Perrow, 1967; Woodward, 1965, discussed in Chapter 4).

Step 2.2: Define Performance Expectations

In this context, the key performance criteria (or categories of performance) related to the organization's purpose and technical processes are identified. First and foremost, this requires a determination of success criteria for products and services, but it may also include performance measures at other points in the organization's system, especially if decision making is important to work process improvement. As described in Kleiner's (1997) framework adapted from Sink and Tuttle (1989), specific standardized performance criteria guide the selection of specific measures that relate to different parts of the work process. Performance measures may be objective or subjective, as in the case of self-reports.

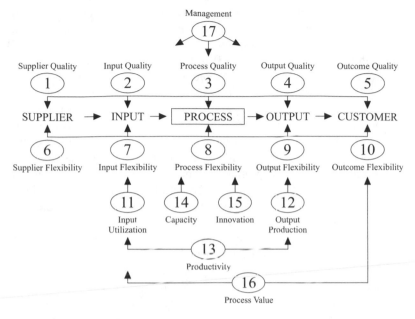

Figure 5.2 Standard checkpoints (adapted from Kleiner, 1997).

Figure 5.2 illustrates the standardized checkpoints or critical points in the work system (Kleiner, 1997). Quality Checkpoints 2 and 4 correspond to traditional measures of quality control, usually assured through inspection of inputs and outputs, respectively (e.g., quality of library materials and produced scholarship). Quality Checkpoints 1, 3, and 5 are quality criteria popularized by Deming and the total quality management (TQM) movement.

Checkpoint 1 emphasizes the quality of suppliers (e.g., authors), which has been operationalized within the quality movement in the form of supplier certification. Checkpoint 3, in-process control, relates to the use of statistical quality control charts to monitor and control processes (e.g., correction of errors identified in posted works). Checkpoint 5 refers to customer satisfaction, operationalized as the customer's (user's) getting what is wanted and needed.

Checkpoint 17 corresponds to total quality management, or the management system by which the other criteria are managed (e.g., management of the library itself). Because of the increasing need to manage and measure flexibility in systems, Kleiner (1997) added a flexibility criterion (Checkpoints 6–10) that relates to each of these checkpoints as well (e.g., ability of the library to change over time). This is especially pertinent to the design of work systems as sociotechnical systems.

The efficiency criterion (Checkpoint 11) focuses on input or resource utilization (e.g., hard drive space). Effectiveness, or the production of output (Checkpoint 12), relates to whether objectives are realized (e.g., goal of increased users). Productivity (Checkpoint 13) is measured as the ratio of outputs to inputs (e.g., scholarship over time). According to Sink and Tuttle (1989), quality of work life includes safety as a criterion (e.g., well-being of librarians). We refer to this checkpoint as the *capacity* (Checkpoint 14) of the system.

Innovation (Checkpoint 15) refers to creative changes to process or product that result in performance gains (e.g., library improvements). Process value (Checkpoint 16) relates to standard business management criteria. For-profit organizations typically use profitability measures to ultimately quantify the "bottom line." For not-for-profit organizations, Sink and Tuttle (1989) introduced *budgetability*, or expenditures relative to budget, to replace the profitability criterion (e.g., library costs vs. revenue).

A popular philosophy in modern organizational performance measurement is to achieve a balanced scorecard, which means managing performance uniformly across the system. In other organizations, it has been seen as useful to identify key performance indicators linked to a few vital strategic goals. In either case, the metrics that correspond to performance criteria should be linked to the organization's strategy.

Step 2.3: Specify Organizational Design Dimensions

Now that the type of production system has been identified and the empirical production models consulted, preliminary levels of complexity, centralization, and formalization can be hypothesized. For example, a digital library, according to Perrow's taxonomy, would be an "engineering" technology. There are well-defined search procedures but a high degree of task variability. Thus, at this juncture, a prescription of moderate centralization with flexibility through low formalization is in order.

Step 2.4: Define System Function Allocation Requirements

From the Clegg et al. (1989) function allocation (i.e., between human and machine) methodology, this is the appropriate time to specify system-level objectives. Requirements specifications can be developed, including micro-ergonomics requirements. Also included are system design preferences for complexity, centralization, and formalization.

Clegg et al. also suggested the use of scenarios that present alternative allocations and associated costs and benefits. In a library work system, the role of the human librarian will vary from the librarian's role in the traditional physical library. The requirements for roles and relationships of human interaction with the system would be defined at this stage.

PHASE 3: Technical Work Process and Unit Operations
3.1 Identify unit operations
3.2 Flowchart the process

Step 3.1: Identify Unit Operations

Unit operations are groupings of conversion steps from inputs to outputs that together form a complete or whole set of tasks and are separated from other steps by territorial, technological, or temporal boundaries. Unit operations often can be identified by their own distinctive subproduct (i.e., component) and typically employ 3–15 workers. They can also be identified by natural breaks in the process (i.e., boundaries determined by state changes or actual changes in the raw material's form, location, or storage.

For each unit operation or department, the purpose/objectives, inputs, transformations, and outputs are defined. If the technology is complex, additional departmentalization (horizontal differentiation) may be necessary. If colocation (physically located together) is not possible or desirable, spatial differentiation (i.e., organization based on different locations) and the use of integrating mechanisms for coordination may be needed. If the task exceeds the allotted schedule, then work groups or shifts may be needed.

Ideally, resources for task performance should be contained within the unit, but interdependencies with other units may complicate matters. In these cases, job rotation, cross training, or relocation may be required. In Figure 5.1, the list under "Transformation Processes" would constitute unit operations.

Step 3.2: Flowchart the Process

The current work flow of the transformation process (i.e., conversion of inputs to outputs) should be flowcharted, including material flows, workstations, and physical and informal or imagined boundaries. In linear systems, such as most production systems, the output of one step is the input of the next. In nonlinear systems, such as many service or knowledge work environments, steps may occur in parallel or may be recursive.

Unit operations are identified at this stage. Also identified are the functions and tasks of the system (Clegg et al., 1989). The purpose of this step is to assess improvement opportunities and coordination problems posed by technical design or the facility. Identifying the work flow before proceeding with a detailed task analysis can provide a meaningful context in which to analyze tasks. Once the current flow is charted, the macroergonomist or analyst can proceed with a task analysis for the work process functions and tasks. A partial flowchart for the publishing process is shown in Figure 5.3.

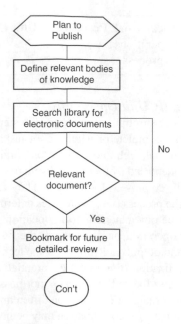

Figure 5.3 Partial flowchart for a digital library.

PHASE 4: Variance Data
4.1 Collect variance data
4.2 Differentiate between input and throughput variances

Step 4.1: Collect Variance Data

A *variance* is an unexpected or unwanted deviation from standard operating conditions, specifications, or norms. As Deming and Shewhart suggested (Walton, 1986), variances can be assigned special causes or common causes, the former being abnormal causes and the latter expected system variation from normal operations. Special or uncommon variances need to be tackled first to get the work process under control. Then, common variation can be tackled for overall system improvement.

For the ergonomist, identifying variances at the process level as well as the task level can add important contextual information for job and task redesign. In a library system, common causes of variation could include typos in works; special causes might be falsified or stolen documents.

Passmore (1988) outlined several principles for design of the technological subsystem which may guide this process. These principles are as follows:

**Principles for Technological Subsystem Design
(adapted from Passmore, 1988)**

1. Variances should be controlled at their source.

2. Boundaries between units should be drawn to facilitate variance control.

3. Feedback systems should be as complex as the variances that need to be controlled.

4. The impact of variances should be isolated in order to reduce the likelihood of total system failure.

5. Technical expertise should be directed to the variances with the greatest potential for system disruption.

6. Technological flexibility should match product variability.

7. Technology should be appropriate to the task.

8. Inputs should be monitored as carefully as are outputs.

9. The core absorbs support (i.e., support functions are the main transformation process).

10. The effectiveness of the whole is more important than the effectiveness of the parts.

By using the flowchart of the current process and the corresponding detailed task analysis, the macroergonomist or analyst can identify variances by comparing what is actually done with what should be done. For example, are authors actually bookmarking and subsequently reading the documents they flag on their initial search, or do they cut and paste the citations for those documents without actually reading them?

Step 4.2: Differentiate between Input and Throughput Variances

Deviations in raw material are called *input variances*. Deviations related to the process itself during normal operations are called *throughput variances*. Both can be identified at this stage by collecting data about the important characteristics of the product as it is developed. Differentiating between types of variances helps determine how to control the variances. An example of an input variance is the variation in format of digital theses and dissertations, depending on the technical domain of the supplier. A throughput variance might be users' browsing techniques.

PHASE 5. Construct Variance Matrix
5.1 Identify relationships among variances
5.2 Identify key variances

Step 5.1: Identify Relationships among Variances

Key variances are those in the tradition of a Pareto analysis, which significantly affect performance criteria and/or may interact with other variances, thereby having a compound effect. The purpose of Step 5.1 is to display the interrelationships among variances in the transformation work process to determine which ones affect which others (see Table 5.2).

The variances should be listed in the order in which they occur in the process down the vertical *y* axis and across the horizontal *x* axis. The unit operations (groupings) can be indicated, and each column represents a single variance. The analyst can read down each column to see if this variance causes other variances by seeing which intersecting cells are checked with an X. Each cell then represents the relationship between two variances. An empty cell implies the two variances are unrelated.

The analyst or team can also estimate the severity of variances by using a Likert-type rating scale. Such a scale might range from "very disruptive" to "not disruptive." Severity would be determined on the basis of whether a variance, or combination of variances, significantly affects performance. This should help identify key variances.

Step 5.2: Identify Key Variances

A *key variance* is a variance that significantly affects quantity of production, operating costs (utilities, raw material, overtime, etc.), or social costs

Table 5.2
Digital Library (DL) Work System Partial Variance Matrix

Unit Operation	Partial Variance Matrix for Digital Library System								
	Digital Library Variance								
Input	1. Social features	1							
	2. Stakeholder roles	X	2						
	3. User motivation	X	Ⓧ	3					
Throughput	4. Multiuser awareness	X			4				
	5. Interoperability				X	5			
	6. Search and retrieval				X		6		
	7. Cognitive models	X	Ⓧ	X	X			7	
	8. Multilingual retrieval	X				X	X		8
	9. Multimedia use			X		X				9

(dissatisfaction, safety, etc.), or has numerous relationships with other variances in the matrix. Typically, consistent with the Pareto principle, only 10%–20% of the variances are significant determinants of the quality, quantity, or cost of product.

In Table 5.2, the circled variances have been identified as key. User motivation is identified as key because the digital library offers a set of tools and services that can be used effectively only if the user has appropriate intentions. Considering the level of flexibility designed into the system, the user with inappropriate intentions could misuse the system.

Related to motivation is the user's cognitive or mental model of the library and his or her role in the library. The system is flexible and fairly decentralized, so the user has significant control in the environment. Although it is assumed that librarians will be available on-line in both synchronous and asynchronous modes, they can only assist the user, not do the work for the user (as in a physical library as well).

PHASE 6: Variance Control Table and Role Network
6.1 Construct key variance control table
6.2 Construct role network
6.3 Evaluate effectiveness
6.4 Specify organizational design dimensions

Step 6.1: Construct Key Variance Control Table

The purpose of this step is to discover how existing variances are controlled and whether personnel responsible for variance control require additional support. The *Key Variance Control Table* includes the unit operation in which variance is controlled or corrected; who is responsible; which control activities are currently undertaken; which interfaces, tools, or technologies are needed to support control; and which communication, information, special skills, or knowledge are needed to support control.

Table 5.3
Key Variance Control Table

Key Variance	Unit Operation	Who Responsible	Control Tasks	Technical Support	Social Support
User Motivation	All	User	Perceptual checking	Pop-up queries	Library guiding principles
Cognitive Model	All	User	Perceptual checking	Pop-up queries	On-line training modules

Step 6.2: Construct Role Network

A *job* is defined by the formal job description, which serves as a contract or agreement between the individual and the organization. This is not the same as a *work role*, which is composed of actual behaviors of a person occupying a position or job in relation to other people. These role behaviors result from actions and expectations of a number of people in a role set. A *role set* comprises people who are sending expectations and reinforcement to the role occupant. *Role analysis* addresses who interacts with whom, about what, and the effectiveness of these relationships. This relates to technical production and is important because it determines the level of work system flexibility.

As illustrated in Figure 5.4, a *role network* is a map of relationships indicating who communicates with the focal role. The work roles are shown as the smaller ovals, and the *focal role* is the larger oval in the center of Figure 5.4. The entire diagram depicts the role network.

Determine focal roles. First, the role responsible for controlling key variances is identified. Although multiple roles may exist which satisfy this criterion, there is often a single role without which the system could not function. The author is identified as the focal role in this user-centered electronic environment because knowledge creation is at the heart of the library's mission.

Draw network. With the focal role identified within a circle, other roles can be identified and placed in the diagram in relation to the focal role. Based on the frequency and importance of a given relationship or interaction, line length can be varied; a shorter line represents more or closer interactions. Finally, arrows can be added to indicate information about the nature of the communication in the interaction. A one-way arrow indicates one-way communica-

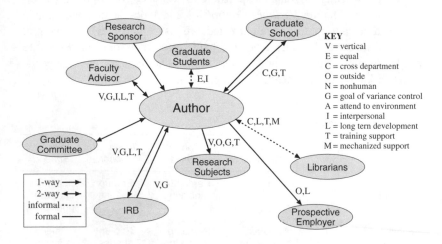

Figure 5.4 Role network for the focal role of Author in a digital library work system.

tion, and a two-way arrow suggests two-way interaction. Two one-way arrows in opposite directions indicate asynchronous (different time) communication patterns.

Characterize interactions. To show the content of the interactions between the focal role and other roles and an evaluation of the presence or absence of a set of functional relationships or functional requirements, the following are typically used to label the role network (see Figure 5.4):

G = short-term *goal* of controlling variances
A = to meet goals, must *adapt* to short-term fluctuations
I = must also *integrate* activities to manage internal conflicts and promote smooth interactions among people and tasks
L = must ensure *long-term development* of knowledge, skills, and motivation in workers

In addition, the presence or absence of particular relationships is determined by describing the work process functions in terms of five types of relationships:

V = *vertical* hierarchy
E = *equals* or peers
C = *cross-boundary*, involving another unit or department
O = *outside* or external stakeholder
N = *nonsocial* component, such as a computer

Step 6.3: Evaluate Effectiveness

The relationships in the role network can now be evaluated by the analyst. Either a Likert-type rating or a simple (+/–) binary system can be used to distinguish between effective and ineffective relationships or interactions. Internal and external customers of roles can be interviewed or surveyed for their perceptions of role effectiveness as well.

Step 6.4: Specify Organizational Design Dimensions

At this juncture, the organizational design hypotheses can be tested against what was learned in the detailed analysis of variance and variance control. The role analysis and variance control table may suggest, for example, a need to increase or decrease formalization or centralization. In a digital library, there would probably be a need to have less formalization because of the high level of professionalism of users, but at the same time, some degree of centralized decision making is needed to promote uniformity and consistency of operation.

If procedures are recommended to help control variances, this increase in formalization must be evaluated against the more general organizational design preferences suggested by the environmental and production system analyses.

PHASE 7: Function Allocation and Joint Design
7.1 Perform function allocation
7.2 Design technological subsystem changes
7.3 Design personnel changes
7.4 Prescribe final organizational design

Step 7.1: Perform Function Allocation

Once system objectives, requirements, and functions are specified, one can systematically allocate functions and tasks to human(s) or machine(s), including computer(s). It is helpful to review the environmental scan data from Phase 1 to check for any subenvironment constraints (e.g., political, financial) before making any mandatory allocations (Clegg et al., 1989). Mandatory allocations are constraints; they must be assigned.

Next, provisional or preliminary allocations can be made to the human(s), machine(s), both, or neither. In the latter case, a return to developing requirements specifications is required. See Kleiner (1998a) for a review of macroergonomics directions and issues in function allocation.

Step 7.2: Design Technological Subsystem Changes

Technological changes are made to prevent or control key variances. Human-centered design of the following may be needed to support operators as they attempt to prevent or control key variances: interfaces, information systems to provide feedback, job aids, process control tools, more flexible technology, redesign of workstations or handling systems, or integrating mechanisms. For example, in a digital library system, users might need access to information about librarian activity at a given point in time. An author in need of assistance would find it useful to access information about where he or she was in the help queue.

Step 7.3: Design Personnel Changes

After considering human-centered system design changes in the previous step, it is time to turn attention to supporting the person directly by addressing knowledge and/or skill requirements of key variances and any selection issues that may be apparent. In developing a variance control table, who controls variances and the tasks performed to control these variances is defined. We recommend personnel system changes to prevent or control key variances. This may entail specific skill or knowledge sets that can be acquired through technical training. The training can consist of formal courses, workshops, or distance learning. For example, specialized training by appropriate law enforcement officials might assist librarians in determining which documents are illegal or falsified. Specialized tools could also assist with this identification.

It is critical to distinguish between communication and information requirements (e.g., Cano, Meredith, & Kleiner, 1998) and to identify specific communication and/or information requirements for performing variance control. This is especially important in virtual work systems, such as a digital library. It may involve support for synchronous (simultaneous) or asynchronous (different time) communication to support information flow. The type of information may vary. For example, the information may be specific and targeted to a particular role or roles, or the information might be shared in a mass communication medium.

Finally, it is important to consider the purpose of the information. Data are collected to be transformed by decision tools (i.e., conversion mechanisms) into information. Information can be shared to increase system awareness/ understanding, or it can be the raw material for decision making, which ultimately should result in positive action in the organization.

Step 7.4: Prescribe Final Organizational Design

At this point in the process, organizational design hypotheses have been generated and adjusted iteratively as new analyses are performed. It is now time to take the specifications for organizational design levels of complexity, centralization, and formalization and produce *specific* structures. Depending on the level of work system process analysis, this may require design/redesign at the organizational level, the group/team level, or both levels. The organizational level is exemplified by choices among functional structures, product structures, geographic structures, or matrix structures. The group level may specify a functional or cross-functional team and a self-managed or high-performance work team. Functional structures (organizing around technical functions such as engineering, marketing, accounting, or production) and functional teams are generally useful in stable environments with smaller workforces (usually classified as fewer than 250 employees).

Functional designs minimize redundancies and provide professional identification and professional advancement. They are often plagued by communication, coordination, and internal competition problems. These problems often result from functional suboptimization in which a department perceives its technical area as the purpose of the organization rather than as serving one or more of the organization's objectives, such as producing goods or services.

Product structures are organized around products/divisions, and geographic structures are organized around regions. These structures represent an attempt to solve the suboptimization problem found in functional structures. However, redundancies in function can often lead to situations in which one function (e.g., payroll) is maintained in multiple organizational units.

The matrix structure is used when considerable flexibility is required in highly dynamic environments. Matrix organizations attempt to exploit the advantages of both the product and functional structures, but they often result in conflicts caused by the existence of dual lines of authority. (See Chapter 4 for more discussion.) Additional integrating mechanisms are often required with matrix designs. These might include a systems engineer or an integrating team.

The cross-functional team typically is used as an integrating mechanism with a functional structure. In cross-functional teams, personnel from different functions form an informal team to focus on issues (e.g., process improvement) that affect the members' functions. The decisions of cross-functional teams result in recommendations to a sponsor with formal organizational authority.

Self-managed teams are formal, in that they are actually part of the organization's overall structure. Rather than make recommendations, self-managed teams have the authority to make decisions about work scheduling, work design, selection/deselection, and so forth.

PHASE 8: Roles and Responsibilities
8.1 Evaluate role and responsibility perceptions
8.2 Provide training support

Step 8.1: Evaluate Role and Responsibility Perceptions

It is important to identify how workers perceive the roles documented in the variance control table, especially if the table was initially constructed by those who do not occupy the roles identified. As exemplified by *Kanzei engineering* in Japan, peoples' perceptions of products and processes are important design considerations (Nagamachi, Matsubara, Nomura, & Sawada, 1996). Through interviews, role occupants can participate in an analysis of their perceptions of their roles.

Using the data collected in Phase 6 and the variance control table, one can identify expected roles, perceived roles, and any gaps between them. Gaps can be managed through training, selection, and technological support. Often in work systems, perceptions of roles will vary, and qualitatively or quantitatively evaluating these perceptual gaps can be instructive to the analyst. For example, in a virtual organization such as a digital library, the extent to which authors actually perceive their roles to be focal might affect overall system effectiveness.

Step 8.2: Provide Training Support

For the ergonomist, this step relates to employee *situational awareness* (how users perceive the system and their role in it) and the existence of a reliable *mental model* (the associated information-processing framework that defines their working understanding of the system and their role in it). In essence, two role networks are operating, the formal one needed for system functioning and the informal one perceived by members. Any variation between the two can be reduced through participatory ergonomics, training, communication, interface design, or tool design. For example, if authors don't perceive the importance of their role in the digital library system, they can be educated about the importance of scholarly works to such a work system.

PHASE 9: Design/Redesign
9.1 Design/redesign support subsystems
9.2 Design/redesign interfaces and functions
9.3 Design/redesign the internal physical environment

Step 9.1: Design/Redesign Support Subsystems

Now that the work process has been analyzed and jointly designed, other internal organizational support subsystems may require redesign. Of particular interest are the information and reward systems. Information systems manage and control information used for decision making in the organization. Reward systems involve how people are paid for their work.

Other subsystems, such as maintenance, may also require adjustment. In all cases, the goal is to determine (a) the extent to which a given subsystem affects the sociotechnical production system, (b) the nature of the variance, (c) the extent to which the variance is controlled, and (d) the extent to which tasks should be taken into account when redesigning operating roles in the supporting subsystem units.

Step 9.2: Design/Redesign Interfaces and Functions

According to the Clegg et al. (1989) method of function allocation, individual and cumulative allocations made on a provisional basis in Step 2.4 can be further evaluated against the following: (a) requirements specification (including the scenarios developed earlier), (b) resources available at the time of implementation (including human and financial), and (c) the total cost of the system (both fixed start-up and continuing). In addition to checking function allocation, interfaces among subsystems should be checked and, as required, redesigned at this juncture.

Step 9.3: Design/Redesign the Internal Physical Environment

Especially at the team and individual levels of work, the internal physical environment – including lighting, humidity, temperature, and so on – should be adjusted if necessary according to ergonomics guidelines to promote effectiveness. Looking at the technical and personnel variance analyses, we can assess whether there are physical environmental changes that can be made to promote improvement. These might include changes to temperature, lighting, humidity, or noise (control/hearing protection, etc).

PHASE 10: Implement, iterate, and improve
10.1 Implement
10.2 Perform evaluations
10.3 Iterate

Step 10.1: Implement

At this point, it is time to implement the work process changes prescribed. In most cases, the macroergonomics team will not have direct authority to implement the changes suggested by the analysis. Therefore, proposals with recommendations for change may have to be prepared for presentation within the formal organizational structure. Such proposals should be consistent with the macroergonomics guiding principles and include both technical (e.g., productivity) and social (e.g., quality of work life) objectives. Proposals will likely include participatory ergonomics as a fundamental methodology.

Any proposal should state expected multidimensional performance improvements (i.e., a combination of performance criteria discussed in Phase 2). Based on the feedback received from proposal presentations, modifications to the proposal may be needed which will require a return to the earlier step, which represents a challenged assumption or design.

Step 10.2: Perform Evaluations

Once the proposal for change is accepted and implementation begins, regular reviews of progress are required. To complement the weekly formative evaluations performed by the implementation team for redesign, efficient semiannual summative evaluations should be performed by an objective outside party. This evaluation should be presented to the implementation team, and a constructive dialogue about expectations and progress to date should be conducted. This process should continue until the team and management determine from the evaluations that the change has been fully implemented and met its objectives. This milestone is usually reached by the end of one year.

Step 10.3: Iterate

As illustrated in Figure 5.5, this process is iterative. For continuous improvement, evaluations may suggest a return to an earlier step for renewed partial or full redesign.

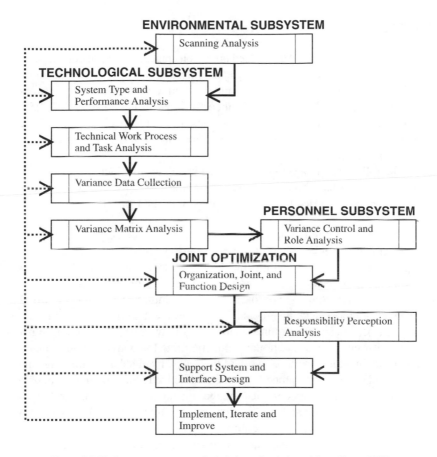

Figure 5.5 Work system process analysis is iterative (adapted from Cano, 1997).

Chapter 6

MACROERGONOMICS RESULTS

Case Studies

In Chapter 2, we noted that, based on sociotechnical systems theory, macro-ergonomics has the potential to greatly improve productivity, scrap rates, safety, health, employee motivation and commitment, and the quality of work life. Specifically, Hendrick (1991) theorized that instead of the 10%–25% improvements in these system effectiveness measures that many of us have experienced from our successful micro-ergonomics interventions, we should see improvements of 60%–90% or more.

The following are a variety of documented cases in which these high improvement rates have been found. (Except for the university college case, these were adapted from Hendrick, 1996, with permission from HFES.)

Red Wing Shoe Company

Beginning in 1985 with initiation of a safety awareness program, a stretching and conditioning program, the hiring of an ergonomics adviser, and specialized training on ergonomics and workstation setup, the Red Wing Shoe Company (Red Wing, Minnesota) made a commitment to reducing work-related musculoskeletal disorders (WMSDs) via ergonomics. The company implemented macroergonomics changes to the work system. Among other things, it went from a traditional production line to continuous flow manufacturing, in which operators worked in groups, received cross training, and rotated jobs. Production processes were modified to reduce cumulative trauma strain. Related micro-ergonomics changes included ergonomically redesigning selected machines and workstations for flexibility, eliminating awkward postures, and greatly easing operation.

As a result of these macro- and micro-ergonomics work system interventions, workers compensation insurance premiums dropped by 70% between 1989 and 1995, for a savings of $3.1 million. The success of this program is attributed to support from upper management, employee education and training, and giving everyone responsibility for coordinating the ergonomics program (Center for Workplace Health Information, 1995b).

AT&T Global

AT&T Global Information Solutions (San Diego, California) employs 800 people and manufactures large mainframe computers. Extensive records and work site analyses were conducted to identify ergonomics deficiencies. Then the company made many micro-ergonomics workstation improvements and provided lifting training for all employees. In a second round of interventions, macroergonomics changes to the work system were implemented. Conveyer systems were replaced with small, individual scissors-lift platforms. The company moved from an assembly-line process to one in which each worker built an entire cabinet and could shift from standing to sitting while at work.

These micro- and macroergonomics work system changes reduced workers compensation costs over the 1990–1994 period by $1.48 million. The cost of implementing these ergonomics work system improvements represented only a fraction of these savings (Center for Workplace Health Information, 1995a).

This case illustrates an ergonomics strategy that we have seen in other organizations. The company starts by making a series of micro-ergonomics improvements that quickly result in positive results (sometimes referred to as "picking the low-hanging fruit"). These results encourage both management and labor to take on more extensive ergonomics projects, which also yield positive results.

Finally, lasting macroergonomics changes to the work system are effected, and these result in the greatest gain. For example, AT&T Global experienced 298 lost workdays due to injury in 1990. This number dropped to 231 the first year after initiating the first (micro-ergonomics) round of changes (1991) and to zero after completing the second (macroergonomics) round of changes (1994 and 1995).

Food Service System Redesign

Through use of a participatory ergonomics approach with food service personnel, two food service stands at Dodger Stadium in Los Angeles were modified. Prior to implementation, two macroergonomists conducted a macroergonomic analysis of the work system and related micro-ergonomics analyses of specific workstation layouts and human-machine interfaces (Imada & Stawowy, 1996). The total cost for the modification project was $40,000. Average customer transaction time was reduced 8 seconds, representing a productivity increase in sales of approximately $1,200 per baseball game. Modifying the other 50 stands in Dodger Stadium is planned at a price of $12,000 per stand.

This modification effort is only the first part of a proposed macroergonomics intervention project to improve the total system process, including packaging, storage, and delivery of food products and supplies and managerial processes (Andrew Imada, personal communication, November 2000).

C-141 Transport Aircraft System: Improving Micro-Ergonomic Design via Macroergonomics

In 1962, Hendrick joined the U.S. Air Force's C-141 aircraft development office as the project engineer for human factors and the alternative mission provisions. The C-141 was to be designed so that its cargo compartment, through the installation of alternative mission kits, could be reconfigured for cargo aerial delivery, paratroop jumping, passenger transport, or medical evacuation. As the plane was initially configured, anything not required for straight cargo carrying was placed in one of the kits, which made them heavy and complex and required considerable time and effort to install.

By reviewing the organizational design and management plan for actual utilization of the aircraft from what, today, we would call a macroergonomics perspective (i.e., analysis of the system's sociotechnical characteristics as described in Chapter 4 and, to some extent, but less formally, as described in Chapter 5), Hendrick was able to identify numerous kit components that rarely ever would be removed from the airplane. The key sociotechnical element necessitating leaving these components in the aircraft at all times was the wide range of missions to be performed, coupled with the high level of environmental uncertainty to which the aircraft had to be prepared to respond rapidly on a global basis.

Using these data, Hendrick worked with the Lockheed design engineers to reconfigure the kits to remove these components and to redesign and install them permanently in the aircraft. This effort greatly simplified the system and reduced actual operational aircraft weight and, thus, related operating and maintenance costs for more than 200 aircraft over a 30-year period. The changes also reduced installation time and labor and storage requirements for the kits. In addition, it saved more than $2.5 million in the initial cost of the aircraft fleet.

We believe this illustrates how macroergonomics considerations can result in highly cost-effective micro-ergonomic design improvements to complex systems.

Petroleum Distribution Company

A macroergonomics analysis and intervention program in a large petroleum distribution company was carried out by Andy Imada. The key components of this intervention included an organizational assessment that generated a strategic plan for improving safety, equipment changes to improve working conditions and enhance safety, and three macroergonomics classes of action items: improving employee involvement and communication and integrating safety into the broader organizational culture.

Imada used a participatory ergonomics approach involving all levels of the division's management and workers. Work system structures and processes were examined from a macroergonomics perspective (much as described in Chapters 4 and 5, but in a less formally structured manner) and, when the analyses indicated a need for change, modified. Employee-initiated ergonomics modifications were made to some of the equipment. New employee-designed safety training methods and structures were implemented. Employees were given a greater role in selecting new tools and equipment related to their jobs.

Two years after installation of the program, industrial injuries were reduced by 54%, motor vehicle accidents by 51%, off-the-job injuries by 84%, and lost workdays by 94%. Four years later, some further reductions were realized (Nagamachi & Imada, 1992). Imada reports that as of 2000, these reductions have largely been sustained and that the company continues to save about $60,000 per year in petroleum delivery costs (Andrew Imada, personal communication, November 2000).

Perhaps the greatest reason for these sustained improvements has been the successful installation of safety as part of the organization's culture (Andrew Imada, personal communication, November 2000). We believe this is a good illustration of how institutionalizing participatory ergonomics within a work system as part of a macroergonomics intervention program can lead to sustained improvements.

Macroergonomics as a TQM Implementation Strategy at L. L. Bean

Rooney, Morency, and Herrick (1993) reported on the use of macroergonomics as an approach and methodology for introducing total quality management (TQM) at the L. L. Bean Corporation, known internationally for the high quality of its clothing products. Using methods similar to those described for Imada's petroleum distribution company intervention, but with TQM as the primary objective, over a 70% reduction in lost-time accidents and injuries was achieved in two years in the company's production and distribution divisions.

Other benefits were also achieved, such as greater employee satisfaction and improvements in additional quality measures. Given the current emphasis in many organizations on implementing ISO 9000, the ISO TQM standard, these results take on even greater significance.

Designing a New University College

All the foregoing cases involved macroergonomics interventions in existing work systems. Hendrick had a unique opportunity to apply macroergonomics in the development of a new, semiautonomous organization, a new university college (Hendrick, 1988). The opportunity occurred in the mid-1980s when a geographically dispersed master's program in systems management was transferred from the University of Southern California (USC) to the University of Denver and was used as the core program for developing a new College of Systems Science. Hendrick transferred with the program for three years to serve as the college's dean during its design and initial development phase.

The systems management program was being taught in university study centers (mini-campuses) at more than 30 locations in the United States and Germany. Hendrick conducted a macroergonomics analysis, as outlined in Chapter 4, with assistance from a special educational technology analysis group from IBM to determine the structure and processes that would be used for the entire work system. Compared with the program as it had existed at USC, this analysis enabled Hendrick to streamline the organizational structure to be more compatible with the college's sociotechnical characteristics (see the description in Hendrick, 1988), improve processes, better design jobs, and make more efficient use of available technology, including computers and software programs.

The college's work system realized a 23% reduction in staffing requirements and about a 25% savings in operating expenses compared with the work system as it had existed at USC. The time required for processing student registrations, grades, and other related administrative activities for the off-campus locations was reduced from an average of three weeks to less than one week. The administrative time demands on the study center managers also decreased approximately 20%, giving them more time to devote to current and prospective students.

Laboratory and Field Research Studies

As we stated in Chapter 1, the subdiscipline of macroergonomics is supported by a long history of empirical research, rooted in the British long-wall mining studies of the 1940s and 1950s. Because random selection and assignment of participants (workers) was not possible in those early investigations, they are more properly classified as "quasi-experiments." Nevertheless, they demonstrated a principle we maintain in modern macroergonomics: To complement case studies, experimentation and quasi-experimentation are needed to build an understanding of what works and why in work system design. Therefore, as we will briefly illustrate, it is both possible and desirable to empirically investigate factors from the personnel, technological, and environmental subsystems and their interactions.

The following discussion is intended to be illustrative, not comprehensive. We aim to demonstrate that sociotechnical factors *can* be investigated in the laboratory and to illustrate how these factors can be manipulated and effects measured. Most of the studies presented here were performed in the Macro-ergonomics Laboratory in the Human Factors Engineering and Ergonomics Center at Virginia Polytechnic Institute and State University (Virginia Tech). In many cases, research problems were identified from actual organizational challenges and brought into the lab for empirical investigation.

Similarly, it is also desirable to return to the field following laboratory investigation for field validation, typically in the form of quasi-experiments, using organizational members and natural work groups as participants.

Quantifying Joint Optimization in Industry

The macroergonomics goal of joint optimization was introduced and defined in Chapter 2. As a theoretical construct, joint optimization makes intuitive sense: To avoid technologically driven systems, the designer or manager should jointly optimize the technology with the personnel. But practically speaking, what does this mean? How does a supervisor interested in joint optimization proceed? How much time should be spent attending to technology versus people? In an initial study, the focus was on understanding the relationship between time allotted to the needs of the personnel and technological subsystems, joint optimization, and department performance. More detail can be found in Grenville (1997) and Grenville and Kleiner (1997a, 1997b).

In this study, 91 full-time first-level supervisors from 12 organizations participated. Each completed a survey that measured the factors necessary to evaluate time allotment to the personnel and technological subsystems. The *level* of joint optimization was measured by each supervisor's perception of 20 critical sociotechnical system state characteristics (Passmore, 1988), such as the extent to which technology supported tasks and the extent to which people were important. Then, time allotment to subsystems was determined by each supervisor's responses to time spent on tasks in the technical and social subsystems, respectively.

Organizational value of time *use* was then estimated by each supervisor's perception of certain constructs related to time, such as the organization's orientation toward scheduling and deadlines, autonomy of time use, awareness of time use, synchronization of tasks, and future orientation (Schriber & Gutek, 1987). Finally, *department performance* was evaluated by scores of the first-level supervisor and plant/warehouse manager on 18 items that estimated performance criteria, such as departmental effectiveness, efficiency, and quality of work life.

There was a strong positive relationship between the level of joint optimization and department performance. Departments with lower joint optimization scores had lower performance scores. Performance scores for outliers (i.e., supervisors with both extreme low and high scores in the data) were compared with scores given by the manager responsible for these supervisors' evaluation. First-level supervisors who split their time 40/60, 50/50, or 60/40 between the technological and personnel subsystems, respectively, tended to score higher on both joint optimization and department performance than did the outliers.

Spending more time on the technological subsystem than on the personnel subsystem (i.e., 60/40 split) tended to result in the highest performance and joint optimization scores. In comparison with other splits, those high-performing first-level supervisors who had a 50/50 time allotment tended to receive performance scores from their evaluating managers that were the most consistent with their own evaluation of departmental performance. This split also yielded the same number of departments with high performance and higher levels of joint optimization as occurred for the 60/40 split (50/50: 10 of 15 managers; 60/40: 10 of 16 managers).

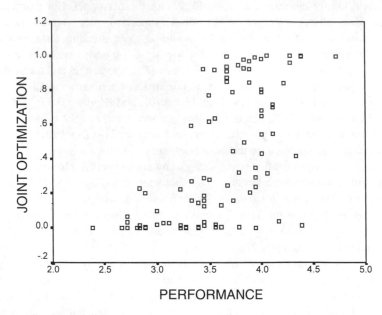

PERFORMANCE

Figure 6.1. Joint optimization versus department performance.

Formalization in Planning Systems in Industry

Planning as a form of organizational decision making is part of the organizational design subsystem of the work system. Planning processes can be formalized to different degrees, and existing planning approaches can be categorized according to their level of formalization. Planning was investigated using a quasi-experimental approach in a classroom environment. Kleiner (1998b) and Jawala (1994) provide more detail on this study.

Planning is the process by which an organization envisions its future and develops the necessary procedures and operations (i.e., decisions) to achieve that future (Nutt & Backoff, 1992). It has traditionally been assumed in strategic planning that an optimal path to the future can be determined for an organization. This typically involves analysis of resources, threats and opportunities, strengths and weaknesses, and the like. However, some scholars have attacked the notion that planning can succeed in a dynamic environment. Because most organizations operate under dynamic environmental conditions, some have challenged the traditional planning methods.

Mintzberg (1994), for example, described the fallacies of predetermination, detachment, and formalization as the explanatory causes for planning system failure. Predetermination is the notion that an organization can predict its future in a dynamic environment. The fallacy of detachment has to do with a traditionally centralized and nonparticipatory planning function. Mintzberg (1994) said that human processes such as intuition and creativity, which are instrumental to planning, cannot be formalized. This study is further detailed in Kleiner (1998b).

In Kleiner's study (1998b), 72 students were randomly assigned to 24 groups of three and were trained in formalized planning; more flexible, informal, and less formalized approach (see Behn, 1988); or a control condition (no specialized training). Following training, groups made organizational decisions in the context of an academic simulation game for eight weeks (simulating four years). The dependent variable was department reputation score, calculated as a function of department productivity, enrollment, faculty job satisfaction related to salary, teaching load, and teaching preference. Each group made decisions about course assignments, salary allocations, promotions, tenure, and hiring.

Data were collected as groups made decisions (game input) and received the results of those decisions (game output). They also received feedback regarding the outcome for the department. Groups made two sets of decisions per academic year. A Likert-type instrument was used to document strategic differences among groups and was completed by each group at the end of each simulated academic year.

The differences in performance among groups were not statistically significant. The hypothesis that formalized planning in a dynamic environment would not yield superior performance was supported. There were significant differences among strategies ($p < .0001$), among groups nested within strategies ($p < .041$), and among years ($p < .001$); there was also a significant Strategy × Year interaction ($p < .026$).

The control groups, which received no specialized strategic training and thus were performing more naturalistic decision making, behaved like low formalization groups. Although the high formalization groups began planning systematically, over time this effect dissipated, likely because of uncertainties in the dynamic environment. This finding suggests that formalization in a dynamic environment may dissipate over time if given the opportunity to do so.

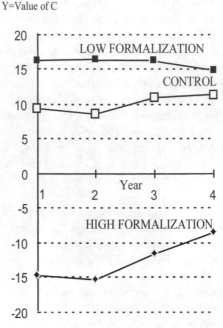

Y=Value of C

Figure 6.2. Results for the three groups (adapted from Kleiner, 1998b). Positive values on the y axis represent a low formalized strategy and negative values represent highly formalized strategies.

Effect of Facilitated Integration on Work Group Performance

A combination laboratory-field experiment explored using a facilitator to perform integration and its effect on personnel processes and performance. These studies demonstrated how macroergonomics laboratory studies can be complemented by field validation in industrial organizations.

Formalized procedures was operationally defined as top-down, imposed, structured procedures required of work groups during an experimental task. Technological subsystem interventions that facilitate group procedures have been shown to affect group performance positively for some tasks (Pavitt, 1993; Pinto, Pinto, & Prescott 1993). The task for the following experiment simulated a process improvement activity that required work groups to generate ideas, select from the ideas, negotiate among group members, and execute group decisions. More information can be found in Hacker (1997).

Students served as participants in the laboratory, and industrial workers participated in the field. The laboratory experiment had 20 student groups, each comprising three members. In the field experiment, the work groups were existing, natural industrial work groups; these were at least twice the size of the student groups. Six industrial organizations provided groups for the field experiment, and 12 groups participated.

The laboratory experiment used a between-subjects design, whereas the field experiment used a within-subjects design. Three mechanisms for formalization of procedures were tested: written procedures, facilitated procedures, and facilitated procedures with technology support (i.e., groupware). In the laboratory, each member role-played either a quality control supervisor, manufacturing supervisor, or purchasing supervisor. Each student was given an objective to achieve corresponding to his or her role.

Control groups experienced low formalization; they were permitted to complete the task any way they chose. The dependent variables included both work group process and performance variables. Work group processes variables were *equality of participation* and *number of ideas brainstormed*. Equality of participation was a directly observable group interaction variable. The sessions were videotaped and analyzed by the researcher to determine levels of equality of participation. In the laboratory study, SYMLOG-A System of Multiple-Level Observation of Groups (Polley, Hare, & Stone 1988) was used to evaluate equality of participation. In the field experiment, equality of participation was measured as the number of comments spoken by each individual.

Work group performance variables were *task performance* and *participant evaluation*. The first measure of task performance was effectiveness. The product had to meet two required specifications. If these specifications were not met, a penalty was assigned during evaluation. The second measure of task performance incorporated several variables. Task performance variables were identified as cost, defect rate, and total task time. Each product produced by a work group was rank ordered according to each measure of task performance (e.g., the lowest-costing product was ranked 1, the second-lowest-cost was ranked 2, etc.). An overall measure of decision quality was used as well, which summed the rank-ordered measures for cost, defect rate, and time for each product.

Facilitated integration did not affect performance but did increase the equality of participation by reducing the overall conversation level within groups in the field. A relationship was found between equality of participation and the overall number of comments spoken in the group ($\rho = -0.75$, $p = .005$). As equality of participation increased, the overall number of comments spoken in the group decreased. The largest change in comments was seen in the most frequently speaking group participant (i.e., identified during the control condition; $p < .0001$). Structure did not affect the number of comments spoken by the least frequently speaking participant, identified during the control condition ($F = 1.94$, $p = .158$). This finding has implications for groups desiring more equal participation among group members but that have a dominant participant in the group who monopolizes interactions.

The findings for this experiment with a task involving multiple-group processes and prior research findings for single-group processes are similar. As shown in Figure 6.3, high integration, even with formalized procedures, produces more ideas than group procedures with medium formalization and low integration (written procedures). Based on the experimental findings, work groups needing to produce a large number of ideas or alternatives should consider using a facilitator for integration. This type of integration increased equality of participation and decreased the overall level of interaction. Facilitated procedures generated more ideas than written procedures alone. Consensus was lower for groups with anonymity evaluating a sensitive issue than with anonymity not present ($p < .0001$).

The findings from this study suggest that level of formalization did not affect groups' decision quality. In addition, the number of comments by members predicted decision quality better than equality of participation. The range in the number of ideas by participants predicted decision quality better than the overall number of ideas of the work group ($R^2 = .858$, $F = 34.16$; $p = .0001$). Anonymity appeared to make a difference in the level of consensus only for sensitive issues. The findings suggest that for sensitive issues, work groups may require anonymity to accurately assess their level of consensus on the issue.

The groupware used in this study allows for anonymous individual voting and ranking (up to 50 users) using individual keypads with summary data

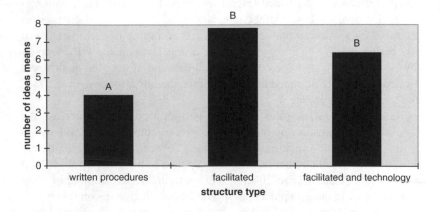

Figure 6.3. Number of ideas generated as a function of type of structure.

displayed to a group. The study in general revealed ways to improve the personnel subsystem through technological intervention and, in terms of research methods, illustrates how laboratory studies with student participants can be validated in the field with actual workers in quasi-experiments and experiments.

Decentralizing Quality Control

Consistent with the decentralization of the quality function, called for by the total quality control movement in the 1960s and reiterated by TQM in the 1980s, the question remains how best to support operators as they accept the responsibility for their own machine and product quality. With the adoption of lean manufacturing and group technology cells in industry, there is the additional issue of how to support an operator as he or she must monitor and control *multiple* machines in a cell.

In Thepvongs and Kleiner (1998), a statistical process control task involved searching control charts for out-of-control conditions and making appropriate decisions. It was hypothesized that the integration of several control charts into one display would minimize decision makers' time to look at separate control chart data, assuming that one operator had to control multiple machines. This integration could come in the typical two-dimensional coplanar display that would combine data from the various systems into one graph, or it could be the result of changing the perspective of the data to 3D. Previous research (e.g., Wickens & Todd, 1990) has found advantages for using both these integrated approaches in different task environments. Tasks in these studies using the 3D perspective were performed faster (scanning and search time) compared with using 2D planar representations of similar data. Additional details about this research may be found in Thepvongs and Kleiner (1998) and Thepvongs (1998).

The experiment manipulated one independent variable, control chart portrayal, at three levels: multiple 2D portrayal, composite 2D portrayal, and composite 3D perspective portrayal. These displays contained quality control chart data from three processes. Type of nonrandom signal was also manipulated: outside the control limits, run, trend, and a phase-related signal that may be attributable to phase-related correlation. The out-of-control events were presented randomly for any given portrayal, although there were two events per out-of-control signal. These out-of-control events were randomly selected from a pool of data sets that contained the required nonrandom signal.

Four data sets for each nonrandom signal were produced, resulting in 16 sets of nonrandom data. Twelve university students volunteered for the experiment. Based on the findings of this research, mental workload (i.e., the NASA Task Load Index; see Wilson & Corlett, 1995) was found to be significantly lower ($F = 9.42, p = .001$) and, as shown in Figure 6.4, decision accuracy was found to be significantly higher ($F = 8.54$, $p = .002$), in the multiple 2D display condition than in either composite display condition. Differences among the displays can be explained by the presence of clutter, overlap, and confusion in both composite conditions, which negatively affected the results in the composite condition. The composite 3D condition also contained depth ambiguity that had negative effects on the objective measures of the study. There were no differences in decision time, search time, or stopping time among the displays.

Generally, this study illustrates the relationship between macroergonomics factors such as operator span of machine control and micro-ergonomics factors such as display design. Although this study demonstrated that increasing the operator's span of control is plausible, it also demonstrated that 3D visualization, though popular, is not always the obvious solution. A poorly designed 3D display can be worse than a less complex 2D display.

Differentiation and Integration in Engineering Design Groups

Increasingly, engineering design is performed by groups of designers rather than by individuals. In addition to questions about the optimal size of a group for engineering design, there is the question of how best to support and integrate the workings of the group through technology use. Technically, concurrent engineering is a popular approach to design, but it has received very little empirical investigation.

Concurrent engineering (CE) is generally recognized as the practice of concurrently designing both the product and its downstream production and

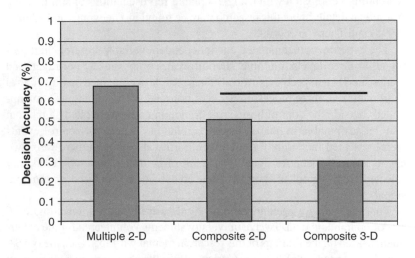

Figure 6.4. Decision making with more automated displays was less accurate.

support processes in the early stages of design (Salomone, 1995). In Meredith (1997), the overall design process methodology was considered – concurrent engineering versus sequential engineering (SE) – and whether to use or not use computer-supported collaborative work technology. Also examined was whether large teams of six persons would be more effective and efficient than small teams of three members.

Determining the number of designers needed to perform the work is a horizontal differentiation issue. Technological support is also a complexity issue because technology serves an integration function. Historically, organizations have used a sequential engineering approach; the product is designed, and then its downstream production and support processes are designed. The use of multidisciplinary or cross-functional teams (e.g., engineering, maintenance, production) is a major tenet of CE (Winner, Pennell, Bertrand, & Slusarczuk, 1988). More information can be found in Meredith (1997).

The research involved a simulated task of designing and producing motorized surface transportation vehicles. The effects of these factors on design performance, process time, process cost, and member satisfaction were determined. In one of the first attempts to simulate concurrent engineering in a laboratory setting, this research did not demonstrate any significant difference between CE and SE. However, a validation field survey of industrial and academic experts in the design process showed unanimous agreement that concurrent engineering shortened product development times, increased product quality, lowered the cost of production, and lowered the life-cycle cost of a product. Clearly, more laboratory research is needed to confirm this popular belief.

The results did, however, reveal an effect of group size on performance. Larger groups were more costly than smaller groups. The hypothesis that larger groups would achieve greater design performance than small groups was not confirmed. Large groups had more ideas to work with, but they were not able to execute those ideas more effectively than were small groups. Larger groups also did not require significantly more process time (109.3 min; i.e., conceptual design, detailed design, manufacturing, or test) than did smaller groups (103.5 min). Large groups were able to employ division-of-labor strategies to maintain efficiency.

Technological support of computer-supported groups cost more than of nonsupported groups ($F = 13.69$, $p = .001$) and did not achieve statistically higher levels of performance. Computer-supported groups had a mean time of 109.50 min, whereas nonsupported groups had a mean time process time of 103.3 min (not significant). This finding is counter to most of the literature on groupware, which suggests that higher efficiency and better decision quality are predicted to be outcomes. However, previous research rarely used engineering design as the task.

This study identified a preferred design group size and suggested that group communication support tools may not be applicable for colocated design work and/or for use when other tools, such as computer-aided design (CAD), are simultaneously used by designers. Further investigation is needed of some of these same factors in a distributed (geographically dispersed) engineering design environment, where communication technologies are required to link group members. Additional details can be found in Meredith (1997).

Integrating Mechanisms for Communication and Decision Making

Increasingly, work systems are virtual or distributed in nature. This study investigated levels of support for the communication process, levels of support for the decision-making process, and levels of support for the sense of presence of distributed (geographically dispersed) groups.

In Cano, Meredith, and Kleiner (1998), the authors tested the effects of high and low levels of decision support on distributed groups. Eight 3-person groups were asked to solve a multicriteria decision problem that involved selecting the best investment alternative among a number of choices under two conditions. In one condition, the groups used a videoconferencing system (presence support) and a communications process software package. In the second condition the groups were asked to solve an equivalent problem, this time using a decision support software package along with the systems used in the first condition.

Results of this study (MANOVA) revealed that in distributed groups with high levels of presence and communication support, higher levels of decision support caused a higher perception of process structure, a higher level of measured consensus, increased decision time, increased decision accuracy, and decreased satisfaction with the group process ($F = 29.127$, $p = .001$). Although some of these results (e.g., increased decision accuracy and increased measured consensus) seemed to be beneficial to the overall group process, others (e.g., increased decision time and decreased satisfaction) are undesirable. The results for decision accuracy and decision time are shown in Figure 6.5.

Some of these results can be explained by the lack of significant effects for perceived consensus and perceived decision quality. It appears that if the decision support system improved the group's decision-making ability, the group members did not perceive such benefits. Under these conditions, the perception of increased process structure seems to decrease the group's level of satisfaction with the process. Implications for work system design include a need to provide group and system performance feedback to members and a need to pay individual attention in design to communication support, decision support, and presence.

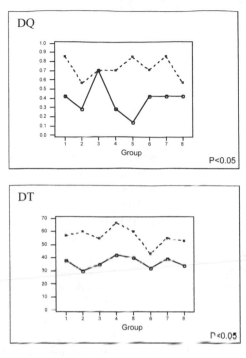

——— Low decision support

------- High decision support

Figure 6.5, Results for decision accuracy (top) and decision time (bottom).

Conclusion

This chapter illustrated how macroergonomics has achieved impressive results in industry, documenting what has worked in various environments and the impressive effects on organizational performance. To complete the circle of knowledge, however, a better understanding of the mechanisms of improvement – how and why they work – is needed. This is what we gain from laboratory and field research. As illustrated here, it is possible to conduct empirical studies of personnel, technological, organizational, and environmental factors and their interactions to derive a fuller understanding of work systems.

Chapter 7

FUTURE DIRECTIONS

Trends Updated

We began this book by highlighting some of the management and organization of work system trends Hal Hendrick identified in 1980 as part of the Human Factors and Ergonomics Society's Select Committee on Human Factors Futures 1980-2000 (Hendrick, 1980, 1991). As we enter the new millennium, it is appropriate to revisit the 1980 trends for an update.

Technology

At the beginning of a new millennium, it is clear that organizations, work groups, and individuals are more dependent on technology than ever. When the technology predictions were proposed in 1980, the personal computer was just being introduced. Today, with networks of PCs, intranets, the internet, smart processes, and smart products, work is increasingly knowledge-dependent – and knowledge increasingly is dependent on technology. We truly live in the age of digitization (Tapscott, 1996).

Perhaps the most challenging of statistics for those attempting to manage technology is that computing power has been doubling every 18 months (Tapscott, 1996, 2000a)! There are significant implications for ergonomists. For example, by 2004, 40% of business-to-consumer e-commerce transactions outside North America will be initiated from portable, wireless-enabled devices (Tapscott, 2000b).

The Industrial Revolution of the 19th century has been replaced by the digital revolution. The conditions are indeed ripe for technology-driven work systems. One need only retrospectively assess the fear, reactivity, and general impact of the concerns over Y2K computer issues. Though it was largely a false alarm, the Y2K hysteria demonstrates the extent to which ours has become a technology-driven society. Information system security breeches, widespread viruses, and concerns about privacy all point to a technology-driven culture.

Thus, more than ever, technology must be jointly optimized with the personnel subsystem. Even with the greatest efforts from many technologists, the human is still a vital component in work systems. Therefore, we believe that systematic work design is imperative. There is a great opportunity for ergonomists equipped with the skills and knowledge of the macroergonomics subdiscipline to serve as work system integrators. In addition to knowledge and skills in macroergonomics, this role will require knowledge and skills in areas such as group dynamics, organizational management, computer science, and human-computer interaction. The macroergonomist who has a broad work system orientation but who is also a specialist in needed technical areas will succeed in the organization of the future.

We live in a virtual world. Human relationships are beginning to depend on technology and electronic communications. Visits or fixed telephone calls to family and friends have been replaced with electronic mail and mobile calls. Although efficiency has increased thanks to such automation, questions arise about the quality of human-human relationships. Without the visual and verbal cues afforded by visual or auditory communication, many relationships have been negatively affected by textual misunderstanding in the electronic communication environment.

In more profound ways, virtual reality (VR) technologies are changing the way we live and work. We have gone beyond the entertainment value of VR; new products are being developed and tested in simulated environments. Surgeons are relying on VR technology to perform diagnostic tests, and robots are being used to perform surgery in some hospitals. Using teleorobotics, coupled with a precision laser beam, doctors are attacking brain tumors without incision, bleeding, or anesthesia. Videoconferencing is being replaced by the 3D representations of meetings in fully immersive environments.

It is clear that as computers become more powerful, their roles in work systems are changing. A major question is, are these new technologies being utilized as human-centered tools to support human endeavor, or are we simply automating that which can be automated? In many cases, implementation is likely a function of feasibility rather than optimality. That is, we automate what we can rather than what makes sense, and humans are allocated to whatever is left over. As many of the research results described earlier suggest, when put to the empirical test, a technology-centered design approach often fails.

In our opinion, macro- and micro-ergonomists must keep pace with these technological changes so that human-centered design and, therefore, human-centered work systems prevail. We cannot be relegated to the system evaluation phase of the design cycle; we should be involved from the beginning. Empirical research must continue to be performed to demonstrate whether the new technologies are indeed useful and, if so, under what conditions. Developments in dynamic and adaptive function allocation must continue.

In 1960, Licklider predicted that by 1980, computers and humans would share symbiotic relationships. In 1980, Hendrick predicted the need for human-centered technology development and related interface design. As we enter the 21st century, we believe the need for a human–centered design ethic is paramount.

Demographic Shifts

The demographic trends identified by Hendrick in 1980 are still valid. In the area of professionalism, the U.S. Bureau of Labor Statistics reported that between 1970 and 1997, the number of people without a high school diploma had decreased from 35% to 11%, and the proportion of the workforce with education beyond the high school level had doubled.

Further, these demographic changes are revealing even greater challenges. The workforce is aging, and work is becoming knowledge-based, requiring the use of information technology (IT). Drucker (1995) warned that the shelf-life for knowledge is no more than five years. To support the increases in professionalism and associated skills and knowledge, the so-called nontraditional student requires IT training and education and, in many cases, must receive distance education. The work systems in which these professionals are involved need to become less formalized than they were when these workers began their careers. These work system design changes, which also include increased decentralization, require macroergonomics intervention. For example, many of the communication and information tools designed to support decision making in decentralized environments are overly formalized.

Thanks to communication and information technology, as well as worldwide competition, work has become increasingly global. This requires an appreciation for cross-cultural issues as well as a new language skill set.

Value Changes

The trends in values identified by Hendrick in 1980 have existed for some time, and there has been a noticeable industry response. Indeed, worker control and decision authority have replaced many of the highly centralized work systems of several decades ago (Drucker, 1995; Macy, 1993). However, because of such factors as the inflated cost of health care and the need to reduce costs quickly, many organizations have downsized and/or gone to the widespread use of temporary or part-time workers. Even so-called permanent workers don't feel a sense of job security and may change jobs frequently. So although the trend has been to "empower" workers, broaden jobs, put workers on teams, and expect their buy-in to organizational mission and vision statements, workers have not perceived a reciprocal level of commitment from their employers. In fact, many organizations have mismanaged these changes (Keidel, 1994). For example, workers have been assigned to teams without proper communication about the expected decision-making authority or reward process. Assuming more authority than they were actually given, many have become disillusioned with teams and associated programs such as TQM. And although it has been customary of late to involve employees in mission and vision sharing, many of them report contradictions in how management behaves.

These kinds of problems make the job of the macroergonomist challenging. He or she must understand the extrinsic and intrinsic motivational needs of the individual, and other personnel characteristics, in order to perform a valid analysis of the personnel subsystem and valid joint design.

World Competition

The competitive characteristics identified in 1980 are still valid, but there is new emphasis on the globalization of corporations and the concomitant integration of dispersed units through a proliferation of communication and information technologies. In many industries, the top 10 competitors will include corporations from several nations (Drucker, 1995). A national company will have international subsidiaries, divisions, and offshore sales and production facilities. Thanks to developments in telecommunications and information technology, time zone differences, geographic distance, language distance, and a host of other factors no longer serve as major barriers to global business. International standards, such as those of the International Standards Organization, and regional trading blocks such as the European Union and the North American Free Trade Agreement have increased world competition as well.

Hendrick also predicted a need for ergonomically sound products and a consideration of the larger organization. With so-called smart products, this need is greater than ever. Smart products, by definition, affect peoples' lives, sometimes in ways not envisioned by their inventors. Beyond the needed interface design for these products, someone needs to attend to the ethical and quality-of-life issues that have emerged.

For the macroergonomist, all this requires a sophisticated understanding of the environmental subsystem and an ability to design environmental interfaces. As discussed earlier, the environmental subsystem is perhaps the most important in terms of organizational performance. The knowledge set required includes socioeconomics, culture, politics, education, and environmental management, among others.

Ergonomics-Based Litigation

The profession and practice of ergonomics have indeed been extended to the courtroom, as Hendrick suggested they would. The liability for injuries and accidents, damages, and the like have been linked to work system design issues – in many cases, attempting to link micro-level design to macro-level performance (including health and safety) and safety and health deficiencies to macro-level design root causes. The ergonomist may be brought in for accident analysis, product analysis, laboratory testing, standards or regulations interpretation, and/or an expert opinion.

As society has become more litigious, ergonomics-based litigation has also increased. This may be partly attributable to the litigious trend in the larger society, but it also may reflect a greater appreciation or awareness of ergonomics. The latter could be traced to the increase in industry and military standards and guidelines related to safety and/or ergonomics. In addition, the explosion in information technology seems to have increased the level of awareness about human-computer interaction issues, including those of work system and job design.

Applicable standards relate to such diverse applications as warning labels, lighting, noise, facility design, computers, and materials handling, The major sources for human factors-related standards and guidelines include the American National Standards Institute, International Standards Organization, Society of Automotive Engineers, Occupational Safety and Health Act, National Institute of Occupational Safety and Health, and the military. OSHA has guidelines that recognize the importance of work system design issues, including management policies, programs, and procedures as they affect safety and health (see OSHA 3123, Occupational Safety and Health Administration, 1993, which is used widely in many industries). The National Safety Council *Accident Prevention Manual* also takes work system design factors into consideration, particularly the volume *Administration and Programs*.

For the macroergonomist, then, knowledge of these standards and guidelines and an appreciation for litigation processes are important. Whether being retained by the law firm representing an injured employee or that of the defending company, the likelihood of being called on to serve as an expert witness is real and is increasing.

Limitations of Traditional (Micro-)Ergonomics

The earlier predictions about the inability of traditional micro-ergonomics, by itself, to fully achieve expected health, safety, quality of work life, and productivity goals have now been realized. As one objectively studies what has not worked and, alternatively, why some interventions do work (as illustrated in the case studies in Chapter 6), one quickly draws the conclusion that a macroergonomics perspective increases the likelihood of significant ergonomics success. When an interface or training program is designed without attention to the larger work system, suboptimization can occur. Similarly, when a technology-centered approach is taken in applying new technology, suboptimization is the likely outcome. A perfectly good micro-ergonomics design could adversely affect, or be affected adversely by, the larger system and its sociotechnical characteristics.

Let us be clear about one point. It is not that micro-ergonomics is no longer valid or needed. Indeed, the explosion of information technology has demonstrated that we need human-machine and human-software interface design more than ever. However, micro-ergonomics needs the perspective and, in many cases, the comprehensive knowledge and skills of macroergonomics to create harmonized systems. Harmonized macro- and micro-ergonomics design can lead to synergistic work system performance improvement and the betterment of individual and societal quality of life.

Other Trends

Change Agent Function

As results like those cited earlier continue to be realized, and as they are publicized, the demand for macroergonomics interventions should grow. One impact of that demand on the human factors/ergonomics profession is likely to be a shift in the role of the HF/E practitioner. Specifically, in addition to serving as part of a health and safety or engineering design team, the macroergonomist is likely to be called on to serve as an organizational change agent consultant to management. With this comes a shift in emphasis away from being a *reactive* technical specialist to being a *proactive* organizational planner and facilitator of work system changes.

In order to carry out this change agent role, HF/E professionals will need to learn additional skills. These include a knowledge of sociotechnical systems theory, macroergonomics, work system design, and organizational theory and behavior and facilitation skills – topics not normally covered in traditional human factors/ergonomics professional education programs, or covered only superficially.

The good news is that this change agent role will enable HF/E practitioners to have a far greater potential impact on improving organizational effectiveness, including safety and health, productivity, and quality of work life.

Macroergonomics as a TQM Strategy

As was demonstrated in the L.L. Bean macroergonomics intervention case cited in Chapter 6, macroergonomics not only integrates well with TQM but also offers what appears to be a highly effective strategy for implementing TQM. Because continuous improvement of the work system is fundamental to TQM, this finding is not surprising.

Further, as Imada demonstrated in the petroleum company case cited in Chapter 6, true macroergonomics intervention improvements tend to be sustained. This appears to happen because a positive change in the corporate culture is an integral outcome of successful macroergonomics interventions.

As macroergonomics becomes better known and understood among organizations, we should see a progressive increase in the number of companies that adopt it as a primary strategy for both implementing and *sustaining* TQM.

Macroergonomics and Supply Chain Management

TQM, at least as a label, is perhaps on its way out. At the time of publication, industry is using *supply chain management* to describe its overall business philosophy and approach. This perspective not only encompasses managing one's own company with TQM types of methods and tools but extends to suppliers and customers. It is characterized by information technology such as *enterprise resource planning* systems. Such technology-intensive, large-scale frameworks have the potential to provide new breakthroughs in efficiency, productivity, and system integration.

However, without macroergonomics to guide joint attention to personnel and technological subsystems, these large-scale frameworks also run the risk of becoming technology-driven systems, leaving "leftover" functions and tasks for humans to perform.

Macroergonomics and Reduction of WMSDs

Perhaps the major single ergonomics occupational health issue in industry has been the high incidence of work-related musculoskeletal disorders (WMSDs). This problem is not new. In the area of manual materials handling, it has been with us since the time ergonomics emerged as an identifiable discipline. With the introduction of visual display terminals in the mid-1970s and their rapid, widespread integration into factories and, especially, offices during the 1980s, there has been a concomitant increase in WMSDs. For example, Bammer (1987) found a close relationship between the introduction of VDTs and initial reporting of WMSDs to a physician by employees at the Australian National University.

Historically, conventional micro-ergonomics interventions to eliminate awkward postures and reduce lifting loads have often proven effective at reducing WMSDs in manual materials handling situations. VDT-related WMSDs have also been reduced somewhat by micro-ergonomics interventions (e.g, by providing furniture and work environments with ergonomics design characteristics) and through ergonomics training on proper workstation layout and adjustment. In both cases, however, WMSDs generally have not been reduced to the levels that should be achievable.

A number of researchers have produced noteworthy work in the area of WMSDs. Several of the cases cited in Chapter 6 demonstrate what is possible in industrial settings when macroergonomics interventions targeted at reducing WMSDs are instituted. Based on these and similar interventions, reductions in WMSDs by 70% or more appear to be achievable via macroergonomics.

With respect to VDT-related WMSDs, Bammer (1990) conducted a meta-analysis of field studies conducted worldwide during the 1980s. The data on biomechanical factors led her to state that (micro-ergonomics) efforts to effect biomechanical improvements were, by themselves, important but insufficient to reduce WMSDs to potentially achievable levels. From her analysis, Bammer concluded that "improvements in work organisation to reduce pressure, and to increase task variety, control, and the ability of employees to work together must be the main focus of prevention and integration." Furthermore, she stated, "ironically, such improvements in work organisation generally also lead to increased productivity" (Bammer, 1993, p. 35).

We believe that using macroergonomics strategies and methods to prevent WMSDs will prove to be a major thrust in the preventive health arena in the early years of the 21st century. As indicated in several of the case studies cited in Chapter 6, we believe that macroergonomics strategies and methods also will be a major thrust for improving industrial safety.

In Summary

Based on the results of macroergonomics interventions in the 1990s–2000s and the trends noted here, the future of macroergonomics should be one of growth in research, methodology, and scope and magnitude of application. To the extent that this proves to be the case, the discipline of human factors/ergonomics should greatly increase in its impact on safety, health, productivity, and the quality of life. Of critical importance will be our professional preparation to take on this important challenge.

REFERENCES

Argyris, C. (1971). *Management and organizational development.* New York: McGraw-Hill.

Bailey, R. W. (1989). *Human performance engineering* (2nd ed). Englewood Cliffs, NJ: Prentice-Hall.

Baitsch, C., & Frei, F. (1984). A case study of worker participation in work redesign: Some suppositions, results and pitfalls. In H. W. Hendrick & O. Brown, Jr. (Eds.), *Human factors in organizational design and management* (pp. 385–393). Amsterdam: North-Holland.

Bammer, G. (1987). VDUs and musculoskeletal problems at the Australian National University – A case study. In B. Knave & P. G. Wideback (Eds), *Work with display units* (pp.270–287). Amsterdam: North-Holland.

Bammer, G. (1990). Review of current knowledge – Musculoskeletal problems. In L. Berlinguet & D. Berthelette (Eds), *Work with display units* (pp.113–120). Amsterdam: North-Holland.

Bammer, G. (1993). Work-related neck and upper limb disorders – Social, organisational, biomechanical and medical aspects. In A. Gontijo & J. de Suza (Eds), *Secundo congresso Lation-Americano e sexto seminario Brasileiro de ergonomia* (pp. 23–38). Florianopolis, Brazil: Ministerio de Trabalho.

Barnard, C. (1938). *Functions of the executive.* Cambridge, MA: Harvard University Press.

Baron, R. A., & Greenberg, J. (1990). *Behavior in organizations: Understanding and managing the human side of work* (3rd ed.). Boston: Allyn & Bacon.

Bedeian, A. G., & Zammuto, R. F. (1991). *Organizations: Theory and design.* Chicago: Dryden.

Behn, R. D. (1988). Management by groping along. *Journal of Policy Analysis and Management, 7,* 643–663.

Bennis, W. G. (1969, July–August). Post bureaucratic leadership. *Transaction,* p. 45.

Brown, O., Jr. (1986). Participatory ergonomics: Historical perspectives, trends and effectiveness of QWL programs. In O. Brown, Jr., & H. W. Hendrick (Eds.), *Human factors in organizational design and management II* (pp. 433–437). Amsterdam: North-Holland.

Brown, O., Jr. (1994). High involvement ergonomics: A new approach to participation. In *Proceedings of the Human Factors Society 38th Annual Meeting* (pp.764–768). Santa Monica, CA: Human Factors and Ergonomics Society.

Brown, O., Jr. (1996). Participatory ergonomics: From participation research to high involvement ergonomics. In O. Brown, Jr., & H. W.Hendrick (Eds.), *Human factors in organizational design and management V* (pp. 187–192). Amsterdam: North-Holland.

Brown, O., Jr., & Hendrick, H. W. (Eds). (1986). *Human factors in organizational design and management II.* Amsterdam: North-Holland.

Burns, T., & Stalker, G. M. (1961). *The management of innovation.* London: Tavistock.

Cano, A. (1997). *Effects of technological support on decision making performance of distributed groups.* Unpublished M.S. thesis, Virginia Polytechnic Institute, Blacksburg, VA.

Cano, A., Meredith, J., & Kleiner, B. M. (1998). Distributed and collocated group communication vs. decision systems support. In P. Vink, E. A. P. Koningsveld, & S. Dhondt (Eds.), *Human factors in organizational design and management VI* (pp. 501–506). Amsterdam: North-Holland.

Carroll, R. (2000, Autumn). The self-management payoff: Making ten years of improvements in one. *National Productivity Review, 19*(4), 61.

Center for Workplace Health Information. (1995a). An ergonomics honor roll: Case studies of results-oriented programs, AT&T Global. *CTD News Special Report: Best Ergonomic Practices*, pp. 4–6.

Center for Workplace Health Information. (1995b). An ergonomics honor roll: Case studies of results-oriented programs, Red Wing Shoes. *CTD News Special Report: Best Ergonomic Practices*, pp. 2–3.

Cherns, A. B., & Davis, L. E. (1975). Goals for enhancing the quality of working life. In L. E. Davis & A. B. Cherns (Eds.), *The quality of working life* (pp. 55–62). New York: Free Press.

Clegg, C., Ravden, S., Corbertt, M., & Johnson, S. (1989). Allocating functions in computer integrated manufacturing: A review and new method. *Behavior and Information Technology*, *8*, 175–190.

Cohen, A. L. (1996). Worker participation. In A. Bhattacharya & J. D. McGlothlin (Eds.), *Occupational ergonomics* (pp. 235–258). New York: Marcel Dekker.

Davis, L. E. (1982). Organizational design. In G. Salvendy (Ed.), *Handbook of industrial engineering* (pp. 2.1.1–2.1.29). New York: Wiley.

DeGreene, K. (1973). *Sociotechnical systems*. Englewood Cliffs, NJ: Prentice-Hall.

Deming, W. E. (1986). *Out of crisis*. Cambridge: MIT Press.

Dess, G. G., Rasheed, A. M., McLaughlin, K. J., & Priem, R. I. (1995). The new corporate architecture. *Academy of Management Executive*, *9*, 7–20.

Dray, S. M. (1985). Macroergonomics in organizations: An introduction. In I. D. Brown, R. Goldsmith, K. Combes, & M. Sinclair (Eds.), *Ergonomics international* (pp. 520–522). London: Taylor & Francis.

Dray, S. M. (1986). The new internationalism in macroergonomics. In O. Brown, Jr., & H. W. Hendrick (Eds.), *Human factors in organizational design and management II* (pp. 499–503). Amsterdam: North-Holland.

Drucker, P. F. (1995). *Managing in a time of great change*. New York: Truman Talley.

Duncan, R. B. (1972). Characteristics of organizational environments and perceived environmental uncertainty. *Administrative Science Quarterly*, *17*, 313–327.

Eason, K. (1988). *Information technology and organizational change*. London: Taylor & Francis.

Emery, F. E., & Trist, E. L. (1960). Sociotechnical systems. In C. W. Churchman & M. Verhulst (Eds.), *Management sciences: Models and techniques* (pp. 83–97). Oxford: Pergamon.

Emery, F. E., & Trist, E. L. (1965). The causal texture of organizational environments. *Human Relations*, *18*, 21–32.

Emery, F. E., & Trist, E. L. (1978). Analytical model for sociotechnical systems. In W. A. Pasmore & J. J. Sherwood (Eds.), *Sociotechnical systems: A sourcebook* (pp. 120–133). LaJolla, CA: University Associates, Inc.

Fitts, P. M. (1951). Engineering psychology in equipment design. In S. S. Stevens (Ed.), *Handbook of experimental psychology* (pp. 365–379). New York: Wiley.

Grenville, N. D. (1997). *A sociotechnical approach to evaluating the effect of managerial time allotment on department performance*. Unpublished masters thesis, Virginia Polytechnic Institute and State University, Blacksburg, VA.

Grenville, N. D., & Kleiner, B. (1997a). Sociotechnical systems approach to time allotment in manufacturing supervision. In *Proceedings of the 6th Annual Industrial Engineering Research Conference* (pp. 709–713). Norcross, GA: Institute of Industrial Engineers.

Grenville, N. D., & Kleiner, B. (1997b). Relationship between sociotechnical joint optimization and perceived department performance in manufacturing organizations. In *Proceeding of the Human Factors and Ergonomics Society 41st Annual Meeting* (pp. 772–776). Santa Monica, CA: Human Factors and Ergonomics Society.

Hacker, M. (1997). *The effect of decision aids on work group performance*. Unpublished Ph.D. dissertation, Virginia Polytechnic Institute and State University, Blacksburg, VA.

Hackman, J. R., & Oldham, G. (1975). Development of the Job Diagnostic Survey. *Journal of Applied Psychology, 60*(2), 159–170.

Hage, J., & Aiken, M. (1969). Routine technology, social structure, and organizational goals. *Administrative Science Quarterly, 12*, 72–91.

Haines, H. M., & Wilson, J. R. (1998). *Development of a framework for participatory ergonomics.* London: Crown.

Hall, R. H., Haas, J. E., & Johnson, N. J. (1967). Organizational size, complexity and formalization. *Administrative Science Quarterly*, June, 303.

Harvey, E. (1968). Technology and the structure of organizations. *American Sociological Review*, April, 247–259.

Harvey, O. J. (1963). System structure, flexibility and creativity. In O. J. Harvey (Ed.), *Experience, structure and adaptability* (pp. 39–65). New York: Springer.

Harvey, O. J., Hunt, D. E., & Schroder, H. M. (1961). *Conceptual systems and personality organization.* New York: Wiley.

Hendrick, H. W. (1979). Differences in group problem solving behavior and effectiveness as a function of abstractness. *Journal of Applied Psychology, 64*, 518–525.

Hendrick, H. W. (1980). *Human factors in management.* Presented at the symposium on "Professional Planning, 1980–2000" at the Human Factors Society 24th Annual Meeting, Los Angeles, CA.

Hendrick, H. W. (1981). Abstractness, conceptual systems, and the functioning of complex organizations. In G. England, A. Negandhi, & B. Wilpert (Eds.), *The functioning of complex organizations* (pp. 25–50). Cambridge, MA: Oelgeschalger, Gunn and Hain.

Hendrick, H. W. (1984). Wagging the tail with the dog: Organizational design considerations in ergonomics. In *Proceedings of the Human Factors Society 28th Annual Meeting* (pp. 899–903). Santa Monica, CA: Human Factors and Ergonomics Society.

Hendrick, H. W. (1986a). Macroergonomics: A conceptual model for integrating human factors with organizational design. In O. Brown, Jr., & H. W. Hendrick (Eds.), *Human factors in organizational design and management II* (pp. 467–478). Amsterdam: North-Holland.

Hendrick, H. W. (1986b). Macroergonomics: A concept whose time has come. *Human Factors Society Bulletin, 30*(2), 1–3.

Hendrick, H. W. (1988). A macroergonomic approach to designing a university college. In *Proceedings of the Human Factors Society 32nd Annual Meeting* (pp. 780–784). Santa Monica, CA: Human Factors and Ergonomics Society.

Hendrick, H. W. (1990). Perceptual accuracy of self and others and leadership status as functions of cognitive complexity. In K. E. Clark & M. B. Clark (Eds.), *Measures of leadership* (pp. 511–520). West Orange, NJ: Leadership Library of America.

Hendrick, H. W. (1991). Human factors in organizational design and management. *Ergonomics, 34*, 743–756.

Hendrick, H. W. (1994). Future directions in macroergonomics. In *Proceedings of the 12th Triennial Congress of the International Ergonomics Association* (vol. 1, pp. 41–43). Toronto: Human Factors Association of Canada/ACE.

Hendrick, H. W. (1995). Future directions in macroergonomics. *Ergonomics, 38*, 1617–1624.

Hendrick, H. W. (1996). *Good ergonomics is good economics.* Santa Monica, CA: Human Factors and Ergonomics Society.

Hendrick, H. W. (1997). Organizational design and macroergonomics. In G. Salvendy (Ed.), *Handbook of human factors and ergonomics* (pp. 594–636). New York: Wiley.

Hendrick, H. W. (1998). Macroergonomics: A systems approach for dramatically improving occupational health and safety. In S. Kumar (Ed.), *Advances in occupational ergonomics and safety 2* (pp. 26–34). Amsterdam: IOS Press.

Hendrick, H. W., & Brown, O., Jr. (Eds.). (1984). *Human factors in organizational design and management.* Amsterdam: North-Holland.

Herbst, P. G. (1975). The product of work is people. In L. E. Davis & A. B. Cherns (Eds.), *The quality of working life* (pp. 439–442). New York: Free Press.

Hickson, D., Pugh, D., & Pheysey, D. (1969). Operations technology and organizational structure: An empirical reappraisal. *Administrative Science Quarterly, 26,* 349–377.

Human Factors and Ergonomics Society. (1998). HFES strategic plan. *Human Factors and Ergonomics Society Directory and Yearbook,* 1998–1999 (p. 388). Santa Monica, CA: Author.

Imada, A. S., Noro, K., & Nagamachi, M. (1986). Participatory ergonomics: Methods for improving individual and organizational effectiveness. In O. Brown, Jr., & H. W. Hendrick (Eds.), *Human factors in organizational design and management II* (pp. 403–406). Amsterdam: North-Holland.

Imada, A. S., & Stawowy, G. (1996). The effects of a participatory ergonomics redesign of food service stands on speed of service in a professional baseball stadium. In O. Brown, Jr., & H. W. Hendrick (Eds.), *Human factors in organizational design and management V* (pp. 203–208). Amsterdam: North-Holland.

Jackson, S. E. (1992). *Diversity in the work place.* New York: Guilford Press.

Jawala, V. (1994). *Measurement and evaluation of alternative planning strategies.* Unpublished master's thesis, Virginia Polytechnic Institute and State University, Blacksburg, VA.

Katz, D., & Kahn, R. L. (1966). *The social psychology of organizations.* New York: Wiley.

Keidel, R. W. (1994). Rethinking organizational design. *Academy of Management Executive, 8*(4), 12–30.

Kleiner, B. M. (1997). An integrative framework for measuring and evaluating information management performance. *International Journal of Computers and Industrial Engineering, 32,* 545–555.

Kleiner, B. M. (1998a). Macroergonomic directions in function allocation. In O. Brown, Jr., & H. W. Hendrick (Eds.), *Human factors in organizational design and management VI* Amsterdam: North-Holland, pp. 635–640.

Kleiner, B. M. (1998b). Macroergonomics analysis of formalization in a dynamic environment. *Applied Ergonomics, 27,* 1–21.

Kleiner, B. M. (1999). Macroergonomic analysis to design for improved safety and quality performance. *International Journal of Occupational Safety and Health, 5,* 217–245. (Special issue)

Lawrence, P. R., & Lorsch, J. W. (1969). *Organization and environment.* Homewood, IL: Irwin.

Licklider, J. C. R. (1960). Man-computer symbiosis. *IRE Transactions on Human Factors in Electronics,* pp. 4–10.

Macy, B. A. (1993). *Research in organizational change and development.* New York: JAI Press.

Magnusen, K. (1970). *Technology and organizational differentiation: A field study of manufacturing corporations.* Unpublished doctoral dissertation, University of Wisconsin, Madison, WI.

McCormick, E. J. (1957). *Human engineering.* New York: McGraw-Hill.

Meredith, J. (1997). *Empirical investigation of sociotechnical issues in engineering design.* Unpublished Ph.D. dissertation, Virginia Polytechnic Institute and State University, Blacksburg, VA.

Meshkati, N. (1986). Major human factors considerations in technology transfer to industrially developing countries: An analysis and proposed model. In O. Brown, Jr., & H. W. Hendrick (Eds.), *Human factors in organizational design and management II* (pp. 351–368). Amsterdam: North-Holland.

Meshkati, N. (1991). Human factors in large-scale technological system's accidents: Three Mile Island, Bhopal and Chernobyl. *Industrial Crisis Quarterly, 5,* 133–154.

Meshkati, N., & Robertson, M. M. (1986). The effects of human factors on the success of technology transfer projects to industrially developing countries: A review of representative case studies. In O. Brown, Jr., & H. W. Hendrick (Eds.), *Human factors in organizational design and management II* (pp. 343–350). Amsterdam: North-Holland.

Mintzberg, H. (1994). *The rise and fall of strategic planning.* New York: Free Press.

Mileti, D. S., Gillespie, D. S., & Haas, J. E. (1977). Size and structure in complex organizations. *Social Forces, 56,* 208–217.

Montanari, J. R. (1976). *An expanded theory of structural determinism: An empirical investigation of the impact of managerial discretion on organizational structure.* Unpublished doctoral dissertation, University of Colorado, Boulder, CO.

Munipov, V. (1990). Human engineering analysis of the Chernobyl accident. In M. Kumashiro & E. D. Megaw (Eds.), *Towards human work: Solutions and problems in occupational health and safety* (pp. 380–386). London: Taylor & Francis.

Nagamachi, M., & Imada, A. S. (1992). A macroergonomic approach for improving safety and work design. In *Proceedings of the 36th Annual Meeting of the Human Factors and Ergonomics Society* (pp. 859–861). Santa Monica, CA: Human Factors and Ergonomics Society.

Nagmachi, M., Matsubara, Y., Nomura, J., & Sawada, K. (1996). Virtual kansei environment and approach to business. In O. Brown, Jr., & H. W. Hendrick (Eds.), *Human factors in organizational design and management V* (pp. 3–6). Amsterdam: North-Holland.

Negandhi, A. R. (1977). A model for analyzing organization in cross cultural settings: A conceptual scheme and some research findings. In A. R. Negandhi, G. W. England, & B. Wilpert (Eds.), *Modern organizational theory* (pp. 285–312). Kent State, OH: University Press.

Nickerson, R. S. (1992). *Looking ahead: Human factors challenges in a changing world.* Hillsdale, NJ: Erlbaum.

Noro, K., & Brown, O., Jr. (1990). *Human factors in organizational design and management III.* Amsterdam: North-Holland.

Noro, K., & Imada, A. (1991). *Participatory ergonomics.* London: Taylor & Francis.

Nutt, P. C., & Backoff, R. W. (1992). *Strategic management of public and third sector organizations.* San Francisco: Jossey-Bass.

Organ, D. W., & Bateman, T. S. (1991). *Organizational behavior.* Homewood, IL: Irwin.

Occupational Safety and Health Administration. (1993). *Ergonomics Program Management Guidelines for Meatpacking Plants* (OSHA 3123). Washington, DC: Author.

Passmore, W. A. (1988). *Designing effective organizations: The sociotechnical systems perspective.* New York: Wiley.

Pavitt, C. (1993). What (little) we know about formal group discussion procedures (A review of relevant research). *Small Group Research, 24,* 217–235.

Perrow, C. (1967). A framework for the comparative analysis of organizations. *American Sociological Review, 32,* 194–208.

Pinto, M., Pinto, J., & Prescott, J. (1993). Antecedents and consequences of project team cross-functional cooperation. *Management Science, 39,* 1281–1297.

Polley, R., Hare, A., & Stone, P. (1988). *The SYMLOG practitioner: Applications of small group research.* New York: Praeger.

Robbins, S. R. (1983). *Organization theory: The structure and design of organizations.* Englewood Cliffs, NJ: Prentice-Hall.

Rooney, E. F., Morency, R. R., & Herrick, D. R. (1993). Macroergonomics and total quality management at L. L. Bean: A case study. In N. R. Neilson & K. Jorgensen (Eds.), *Advances in industrial ergonomics and safety V* (pp. 493–498). London: Taylor & Francis.

Salomone, T. A. (1995). *What every engineer should know about concurrent engineering.* New York: Marcel Dekker.

Schriber, J. B., & Gutek, B. A. (1987). Some time dimensions of work: Measurement of an underlying aspect of organization culture. *Journal of Applied Psychology, 72,* 642–650.

Sink, D. S., & Tuttle, T. C. (1989). *Planning and measurement in your organization of the future.* Norcross, GA: Industrial Engineering and Management Press.

Smith, A. (1970). *The wealth of nations.* London: Penguin. (Originally published 1876)

Snow, M., Kies, J., & Williges, R. (1997). A case study in participatory design. *Ergonomics in Design, 4,* 18–24.

Stevenson, W. B. (1993). Organizational design. In R. T. Golembiewski (Ed.), *Handbook of organizational behavior* (pp. 141–168). New York: Marcel Dekker.

Sullivan, L. (1999). Tomorrow's workplace. *Risk Management,* 1999, September, *46*(9), 9.

Szilagyi, A. D., Jr., & Wallace, M. J., Jr. (1990). *Organizational behavior and performance* (5th ed.). Glenview, IL: Scott Foresman.

Tapscott, D. (1996). *The digital economy.* New York: McGraw-Hill.

Tapscott, D. (2000a). Transmeta chip may hold key for computing's future. *Computerworld, 34,* 36–38.

Tapscott, D. (2000b). It's beginning to look a lot like a wireless world. *Computerworld,* 34, 32–34.

Taylor, F. W. (1911). *Principles of scientific management.* New York: Harper.

Thepvongs, S. (1998). *Use of integrated process control displays in work system design.* Unpublished master's thesis, Virginia Polytechnic Institute and State University, Blacksburg, VA.

Thepvongs, S., & Kleiner, B. M. (1998). Inspection in process control. In *Proceedings of the Human Factors and Ergonomics Society 42nd Annual Meeting* (pp. 1170–1174). Santa Monica, CA: Human Factors and Ergonomics Society.

Thomas, R. R., Jr. (1991). *Beyond race and gender.* New York: AMACOM.

Thompson, J. D. (1967). *Organizations in action.* New York: McGraw-Hill.

Trist, E. L., & Bamforth, K. W. (1951). Some social and psychological consequences of the longwall method of coal-getting. *Human Relations, 4,* 3–38.

Trist, E. L., Higgin, G. W., Murray, H., & Pollock, A. B. (1963). *Organizational choice.* London: Tavistock.

Van Aken, E. M., & Kleiner, B. M. (1997). Determinants of effectiveness for cross-functional design teams. *Quality Management Journal, 4,* 51–79.

Van de Van, A. H., & Delbecq, A. L. (1979). A task contingent model of work-unit structure. *Administrative Science Quarterly, 24,* 183–197.

Walton, M. (1986). *The Deming management method.* New York: Putnam.

Weber, M. (1946). *Essays on sociology* (trans. H. H. Grath & C. W. Mills). New York: Oxford.

Wickens, C. D., & Todd, S. (1990). Three-dimensional display technology for aerospace and visualization. In *Proceedings of the Human Factors Society 34th Annual Meeting* (pp.1479–1483). Santa Monica, CA: Human Factors and Ergonomics Society.

Wilson, J. R. (1995). *Ergonomics and participation.* In J. R. Wilson & E. N. Corlett (Eds.), *Evaluation of human work* (pp.1071–1096). London: Taylor & Francis.

Wilson, J. R., & Corlett, E. N. (1995). *Evaluation of human work.* London: Taylor & Francis.

Wilson, J. R., & Haines, H. M. (1997). Participatory ergonomics. In G. Salvendy (Ed.), *Handbook of human factors* (pp. 490–513). New York: Wiley.

Winner, R. I., Pennell, J. P., Bertrand, H. E., & Slusarczuk, M. M. G. (1988). *The role of concurrent engineering in weapons system acquisition* (Final R-338). Alexandria, VA: Institute for Defense Analyses.

Woodward, J. (1965). *Industrial organization: Theory and practice.* London: Oxford University Press.

Yankelovich, D. (1979). *Work values and the new breed.* New York: Van Nostrand Reinhold.

Yankelovich, D. (1988). *Starting with the people.* Boston: Houghton Mifflin.

Zwerman, W. L. (1970). *New perspectives on organization theory.* Westport, CT: Greenwood.

GLOSSARY

Abstract functioning: Cognitively complex conceptual functioning characterized by a high degree of differentiation and integration.

Adhocracy: A rapidly changing adaptive work-system organized around problems to be solved by groups of relative strangers with diverse professional skills.

Advanced information technologies (AIT): Emerging computer-based innovations in the early stages of their life cycle that make use of such capabilities as broadband and wireless.

Agile manufacturing: A manufacturing philosophy and methodology that enables a firm to respond successfully to changes in the marketplace.

Bottom-up: A work system analysis and design approach that proceeds from the individual worker level up through the work system's subunits to the overall work system level.

Boundaries: Work system borders that separate domains of responsibility.

Centralization: The degree to which formal decision making is concentrated in a relatively few individuals, group, or level, usually high in the organization.

Change agent: A person whose role is to facilitate change within an organization to better enable it to meet its goals.

Checkpoint: Specific, standardized places in the work process of a work system.

Cognitive complexity: A higher-order structural personality trait; the extent to which people have developed differentiation and integration in their conceptual functioning.

Cognitive ergonomics: The aspect of ergonomics concerned with the design of the interfaces between human mental, perceptual, and information processing characteristics with other sociotechnical system elements – particularly software.

Collaboration: Group participation on a common task.

Complexity: The degree of *differentiation* and *integration* existing in a work system.

Compatibility: The sociotechnical principle that for a work system to exhibit certain characteristics (e.g., human-centered), its design or redesign process must incorporate those same characteristics.

Computer-aided design (CAD): Designing products using computers; utilizes computer software programs specifically developed for aiding in design.

Computer-integrated manufacturing (CIM): Linking together through computer technology all the various departments in an industrial company so they operate smoothly as a single, integrated business system.

Computer-supported collaborative work technology: Information technology that supports human-to-human interaction for shared tasks.

Concrete functioning: Cognitively simple conceptual functioning, characterized by a low degree of differentiation and integration.

Concurrent engineering: Engineering functions, such as design, analysis, and production, that are performed simultaneously with a lot of cross-functional interaction.

Consensus: Final decision agreement/support within a group; does not necessarily mean that individual members agree personally.

Consumer feedback: Information from those who received products and/or services about their satisfaction with those products and/or services.

Consumers: Those who receive and usually pay for outputs (i.e., products and/or services) from a work system.

Continuous flow manufacturing: Similar to *just-in-time* and *flexible manufacturing;* the primary objective is to produce a high-quality product in the shortest possible production time at the lowest possible cost. Provides techniques to reduce product cycle times, minimize inventories, improve quality, and increase inventory turns.

Continuous improvement: The incremental betterment of performance over time. In the sociotechnical systems literature, it is sometimes called *incompletion.*

Control chart: A quality assurance tool that uses historical data to create upper and lower control limits (from standard deviations) to plot data and evaluate current performance of processes.

Coupling: Whether participative data are directly or remotely used, in which the former involves little or no filtering of participant input and the latter involves some filtering or translation, usually by managers or consultants.

Cross-functional teams: A parallel or informal team composed of members from several functional units or departments. These members can represent different skills and ranks. The team usually has partial decision-making authority (i.e., recommendations) and often is involved in TQM (see *Total quality management*) or other process improvement activities.

Decision support technologies: Mechanisms that assist in converting data to information so decisions are easier or better.

Deming flow diagram: A type of organizational input/output diagram popularized by the late W. Edwards Deming, beginning with his involvement in Japan in the 1950s and extending through the TQM movement in the United States in the 1980s.

Departmentalization: Division of a work system's labor into groups of specialists.

Differentiation: The number of conceptual categories a person has developed for storing experiential information. In organizational design, the number of hierarchical levels and/or departments that comprise the structure of the work system.

Domain: An organization's range of products or services offered and market share.

Empowered: Having a sufficient degree of decision-making authority.

Environmental uncertainty: The extent to which an organization's specific task environment is (a) complex, in terms of its number of components, and (b) dynamic versus stable over time.

External environment: Factors external to the organization that permeate it. Examples are materials sources, customers, government policies and regulations, and stockholders.

Facilitator: A person who manages the meeting process of a group; for example, managing the time and uniformity of participation.

Fitt's list: Original list of basic human versus machine capabilities and limitations to guide system design function allocation.

Flexible manufacturing: Ability for the internal manufacturing system to cope with changes dictated by a dynamic environment.

Flexibility: The capability to change in response to environmental change.

Focal role: The work system function that is expected to control the most significant key variances.

Focus groups: A temporary collection of selected participants to discuss specific issues or test market-specific ideas or products.

Formalization: The extent to which jobs within a work system are standardized, including use of formal rules and procedures and explicit job descriptions.

Free-form design: An adhocracy type of organizational design in which there is no functional departmentalization and the shape of the work system changes rapidly in response to its external environment in order to survive. Departmentalization is replaced by a profit center arrangement.

Full direct participation: Having all those affected by a decision or design become involved.

Function allocation: A methodology for assigning tasks and/or functions to humans and/or machines.

Function analysis modeling: A macroergonomics approach for identifying work system functions and related quantitative and qualitative personnel subsystem requirements.

Gap: A variance or deviation from what is expected or needed.

Garbage can model for organizational design: A modification of the garbage can model for organizational decision making for use as a tool to evaluate work system design alternatives.

Harmonized work system: A work system in which all subsystems and components are synchronized and behave as a single unit.

High involvement: On the upper end of the participation continuum, exemplifying significant design or decision-making involvement.

Horizontal differentiation: The degree of departmentalization and specialization within a work system.

Human-centered approach: An approach to human-machine function and task allocation that first considers the capabilities and limitations of the human and whether the function or task justifies the use of a human. Also called the *humanized task approach.*

Human relations theory: In contrast to Classical theory, a school of thought that centers on the assumption that workers' feelings and attitudes are important and can possibly have an impact on performance.

Human-system interface technology: The unique technology of the human factors/ergonomics discipline; it consists of empirically derived design principles, guidelines, specifications, tools, and methods to design human-organization, human-job, human-machine, human-software, and human-environment interfaces.

Ideal bureaucracy: The classical bureaucratic design developed by Max Weber.

Incompletion: See *Continuous improvement.*

Input variance: An unexpected or unwanted deviation of a resource from standard operating conditions, specifications, or norms.

Integration: The number of rules and combinations of rules a person has developed for integrating conceptual information. In organizational design, the number of mechanisms designed into the work system for ensuring communication, coordination, and control among the differentiated elements (e.g., standard operating rules and procedures, committees, task teams).

Internal control: Mechanisms and/or processes that serve as checks and balances on internal processes in a work system.

Job: A formal position in an organization as documented and detailed by a formal job description.

Job enlargement: Tasks added at the same level of responsibility.

Job enrichment: Tasks added at a higher level of responsibility.

Joint design: Attending to both personnel and technical factors simultaneously in the design process.

Joint optimization: A sociotechnical systems design principle that states that the technological and personnel subsystems must be designed jointly in order to achieve the most effective functioning of the work system.

Kansei engineering: Ergonomics technology of product development that translates a consumer's feelings about a new product into design requirements.

Key variance: A deviation from what is expected, needed, or wanted that affects key performance criteria significantly or a variance that has a multiplicative relationship with other variances.

Key variance control table: A tabular representation of which roles control which key variances and how. In addition, special technological support and training requirements needed to control key variances are identified.

Knowledge-based technology: A means of classifying technology based on task variability, or the number of exceptions or nonroutine problems created by the technology, and task analyzability, or extent to which the problems created lend themselves to rational-logical, quantitative, and analytical thinking as opposed to reliance on the experience, judgment, and intuition of the problem solver.

Likert-type (survey): A questionnaire with a graded response to each statement, typically on a 5-point scale: Strongly agree, agree, undecided, disagree, and strongly disagree.

Machine bureaucracy: The bureaucratic form that evolved from Max Weber's *ideal bureaucracy* and Fredrick W. Taylor's *scientific management.* It is characterized by narrowly defined jobs, routine and well-defined tasks, a well-defined hierarchy, high formalization, and centralized decision making.

Macroergonomic analysis and design (MEAD): A 10-step framework for conducting work system improvements.

Macro-ergonomic level: The overall work system level of ergonomics application.

Macroergonomics: The subdiscipline of ergonomics that focuses on the design of the overall work system. Conceptually, a top-down sociotechnical systems approach to the design of work systems and the carry-through of the overall work system design characteristics to the micro-ergonomic design of human-job, human-machine, and human-software interfaces to ensure that the entire work system is fully harmonized.

Mass production: The mode of production in which the items produced are essentially the same and thus lend themselves to mass production techniques, such as assembly lines.

Matrix organization: An adhocracy type of organizational design that combines departmentalization by function with departmentalization by project or product line.

Mechanistic work systems: Work systems characterized by high vertical and horizontal differentiation, formalization, and centralization. They typically have routine tasks and programmed behaviors and can respond to change only slowly.

Mental model: Cognitive representations based on past experience that guide current perceptual activity.

Micro-ergonomics: Those aspects of ergonomics primarily focused on the design of the interfaces between the individual and other system elements, including human-job, human-machine, human-software, and human-environment interfaces.

Middle-out: A work system analysis and design approach that proceeds from an intermediate or subunit level of the work system both up to the overall work system level and down to the individual worker level.

Mission: The goal or purpose of a work system.

Modular form: A relatively new form of adhocracy that outsources nonvital functions while retaining full strategic control.

Monte Carlo technique: An empirical study of statistics using random numbers. It is applied to empirical studies of behavioral models or methods that the investigator wishes to explore. A way of randomly selecting a variety of items within a set so the selection is constrained by the correlations between the dimensions.

Open systems: Systems that are open to being influenced by, and influencing, their external environment, such as sociotechnical systems.

Organic work systems: Work systems characterized by relatively low vertical differentiation and formalization with decentralized tactical decision making, enabling them to be flexible and adapt quickly to change.

Organization: The planned coordination of two or more people who, functioning on a relatively continuous basis and through division of labor and a hierarchy of authority, seek to achieve a common goal or set of goals.

Organization design: The design of a work system's structure and related processes to achieve the organization's goals.

Organizational requirements definition tools (ORDIT): A set of automated tools designed to assist in the specification of work system requirements for information technology systems using an integrated methodology. Developed by the HUSAT Research Institute at the Loughborough University of Technology, UK.

Pareto analysis: Any procedure that identifies which 20% of the variance causes 80% of an impact on performance.

Partial direct participation: Having a representative subset of participants involved, rather than the entire group, usually necessary because of economic considerations.

Participation: A general term for involvement of users or others in a task.

Participative management: A style of supervision that involves workers in decision making, at least at the level of providing recommendations and sometimes at a level of complete delegation.

Participatory ergonomics: The involvement of employees in the ergonomic analysis and design of their work environments and activities.

Passive-aggressive: The acting out of anger or hostility by being passive, not doing things, or doing them slowly or inefficiently. In organizations, this often takes the form of doing the minimum to get by rather than what is really required to get the job done effectively.

Personnel subsystem: One of the four basic elements of a sociotechnical system; consists of the people who make up the organization's workforce.

"Picking the low-hanging fruit": Selecting projects for ergonomic intervention where there is a high probability of showing improvement in productivity, health, safety, or other important organizational criteria in a relatively short time.

Presence support: Technological augmentation that attempts to simulate the perception or feeling of being in a real environment.

Principles: The guiding values (for behavior) of a work system.

Process: A series of steps that convert inputs to outputs in a system.

Process production: The mode of production in which the production process is continuous, such as oil and chemical refineries.

Production technology: A means of classifying technology based on mode of production (e.g., unit, mass, or process).

Production type. The categories of production systems offered by various taxonomies such as "craft," "unit," "mass," and "process" in the case of the production mode classification system.

Professional bureaucracy: A bureaucratic design that relies on a high degree of professionalism in the jobs that make up the work system. It is characterized by more broadly defined and less routine jobs than found in a machine bureaucracy, relatively low formalization, and decentralization of tactical decision making.

Profit center: A structural characteristic of free-form adhocracies used in place of departmentalization. Profit centers consist of highly professionalized, results-oriented work teams.

Psychosocial: The reciprocal influence or interaction of the mental and emotional characteristics of the individual with the social characteristics of the group.

Quality circles: Permanent organizational teams focused on problem solving, popularized by the Japanese in the 1960s.

Quasi-experiment: An empirical study in which independent variables are manipulated to evaluate the effect on dependent variable(s) but in which representative selection and/or assignment are not performed.

Re-engineering: The redesign of work processes to improve efficiency and productivity.

Relevant task environments: Those parts of an organization's external environment that can positively or negatively influence the organization's effectiveness.

Role analysis: The process of evaluating work roles. This can include the expectations others have for these roles as well as the perceptions held by role occupants.

Role network: Sometimes called *role set,* the conceptualization of various work roles and how they interact with one another to form a social subsystem.

Role set: Sometimes called *role network,* the conceptualization of various work roles and how they interact with one another to form a social subsystem.

Scanning: A term from sociotechnical systems theory referring to a broad-based evaluation or analysis, usually focused on the system or environment being studied or improved. A preliminary high-level analysis.

Scientific management: A method of work design, developed by Frederick W. Taylor at the beginning of the 20th century, that involved the systematic observation of workers to determine the "one best way" of performing each task and then training workers to follow it.

Self-managed teams: In contrast to a cross-function team, a permanent team with significant decision-making authority, usually including selection, assignment, scheduling, and work design. These teams are sometimes called *autonomous* or *high-performance* work teams.

Semistructured interview: An interview procedure in which a set of basic questions are developed to be asked to each interviewee, but then the interviewer improvises with additional questions as required to follow up the interviewee's answers to the basic questions to gain additional information.

Sequential engineering: Engineering functions, such as design, analysis, and production, are performed independently with the output of one function serving as the input to the next.

Situational awareness: Perception of the immediate environment.

"Smart" products: System outputs with ingrained information-processing intelligence.

Social boundaries: The borders created by the formal organization chart's definition of jobs.

Sociotechnical systems: Work systems composed of (a) a technological subsystem, (b) a personnel subsystem, (c) an external environment that interacts with the organization, and (d) an organizational design.

Span of control: The number of employees a given manager can directly supervise effectively. Span of control is affected by a number of work system design factors.

Spatial dispersion: The extent to which an organization's activities are performed in multiple locations. It is measured by (a) the number of geographic locations constituting the total work system, (b) the average distance of the separated locations from the organization's headquarters, and (c) the proportion of employees in these separated units relative to the number in the headquarters.

Specific task environment: The particular combination of relevant task environments for a given organization.

Stakeholders: Individuals or groups with a vested interest in a work system.

Steady state: A constant trend in performance.

Strategic decisions: Decisions that deal with the long-range vision and goals of the organization.

Strategic planning: Long-range planning for an organization, usually looking 5-10 or more years into the future. Strategic planning typically results in strategic planning documents and "roadmaps." Popularized by General Electric in the 1960s.

Stratified semistructured interview: A procedure in which a sample of interviewees is systematically selected by organizational level, department, and so on to ensure that it will be representative of the entire work system of interest. Each selected interviewee then goes through a semistructured interview process with an interviewer. This procedure is widely used in conducting organizational assessments.

Structural analysis: Analysis of the organizational structure of a work system.

Structural form: The type of organizational structure utilized by a given work system.

Subenvironments: Categorical subunits within the external environment of a work system.

Suppliers: Providers of resources or inputs to a work system.

Survey feedback method: A method in which the data gained from an organizational questionnaire survey are summarized and subgrouped statistically and by organizational level, department, project, and so on and then are fed back to the individual organizational units for interpretation and, where applicable, action to improve organizational functioning.

System harmonization: The work system condition achieved when all subsystems are synchronized and behaving as a single unit.

System inputs: Resources provided by suppliers to a work system.

Systematic organizational design methodology (SORD): Developed for designing U.S. Army organizational units, SORD is a step-by-step set of computer-assisted procedures for the comprehensive, systematic, integrative, and reliable design of work systems.

Tactical decisions: Decisions that deal with the day-to-day operation of the work system.

Task allocation: The process of assigning tasks to humans or machines in designing or modifying a sociotechnical system; includes the allocation of tasks to specific work modules and jobs.

Technological complexity: A scale of production mode complexity, with unit production being the least technologically complex, mass production intermediate, and process production the most complex.

Technological imperative: The often-held view, unsupported by the research literature, that technology has a compelling influence on work system structure and should determine work system design.

Technological subsystem: One of the four basic elements of a sociotechnical system; consists of the machines, tools, software, and other technological components of the organization.

Territorial boundaries: The borders around the physical space used for product conversion.

Throughput boundaries: The work system borders, from the input owned by the system to the output for distribution to consumers.

Throughput variance: An unexpected or unwanted deviation in a process from standard operating conditions, specifications, or norms.

Time boundaries: The temporal borders related to such characteristics as seasonality and number and timing of shifts.

Top-down: A work system analysis and design approach that proceeds from the overall work system level down through the work system's subunits to the individual worker level.

Total quality management (TQM): A management system and philosophy that uses cross-functional teams and tools to continuously improve business processes for the satisfaction of customers. Deming, Juran, and Crosby are some of the names associated with the movement. In the United States, the Malcolm Baldrige National Quality Award recognizes companies for achievement in this area.

Unit operations: Groupings of conversion steps from inputs to outputs that together form a complete or whole set of tasks and are separated from other steps by territorial, technological, or temporal boundaries.

Unit production: The mode of production in which each item produced is unique, rather than essentially the same; for this reason, items do not lend themselves to mass production techniques.

Usability: The extent to which a given hardware or software product can readily, effectively, and safely be operated or maintained (used) by people from the intended user population; often dependent on the extent to which the product has been well designed ergonomically.

User-centered design: A design philosophy that uses participative approaches to involve users in the design process.

User systems analysis: An approach for assessing work system needs for information-processing equipment and software and evaluating related task and work system design factors. Central to the approach is an analysis of the user's environment, functions, and tasks, and related user information needs.

Value: An attribute that guides behaviors and attitudes.

Variance: An unexpected or unwanted deviation from standard operating conditions, specifications, or norms.

Variance table: A graphic representation that positions variances along both axes and illustrates which variances are related to which other variances. Key variances are usually identified as well.

Vertical differentiation: The hierarchical structure of the work system. It is measured by the number of hierarchical levels separating the chief executive position from the jobs directly involved with the work system's output.

Virtual organization: A relatively new form of adhocracy that consists of a continually evolving network of independent companies.

Virtuality: The simulation of real environments or tasks through information technologies.

Vision: The long-term view or desire of what a work system is to become in the future.

"Walk the talk": Managers exhibiting the same behaviors and/or attitudes (driven by corporate values) that they expect of their subordinates.

Work flow integration: A scale for defining technology in both manufacturing and service organizations in terms of a combination of three factors: (a) degree of equipment automation, or extent to which work activities are performed by machines; (b) work flow rigidity, or the extent to which the sequence of activities is inflexible; and (c) specificity of evaluation, or the degree to which work activities can be assessed by specific, quantitative means.

Work-related musculoskeletal disorders (WMSDs): Musculoskeletal disorders that result from work that requires excessive lifting, repetition, awkward postures, or other stress factors in the work environment, including psychosocial factors.

Work role: The actual function and tasks expected or required to be performed to control variances in a work system. These functions and tasks may or may not be consistent with the formal job.

Work system: A system that involves two or more persons interacting with some form of (a) hardware and/or software, (b) internal organizational environment, (c) external environment, and (d) organizational design.

INDEX